**ACPL ITEM
DISCARDED**

1-10-75

Coal Gasification

Coal Gasification

Lester G. Massey, *Editor*

A symposium sponsored by
the Division of Fuel
Chemistry at the 165th
Meeting of the
American Chemical Society
Dallas, Texas,
April 9-10, 1973.

ADVANCES IN CHEMISTRY SERIES **131**

AMERICAN CHEMICAL SOCIETY
WASHINGTON, D. C. 1974

ADCSAJ 131 1-266 (1974)

Copyright © 1974

American Chemical Society

All Rights Reserved

Library of Congress Catalog Card 74-78009

ISBN 8412-0195-1

PRINTED IN THE UNITED STATES OF AMERICA

Advances in Chemistry Series
Robert F. Gould, *Editor*

Advisory Board

Kenneth B. Bischoff

Bernard D. Blaustein

Ellis K. Fields

Edith M. Flanigen

Jesse C. H. Hwa

Phillip C. Kearney

Egon Matijević

Thomas J. Murphy

Robert W. Parry

FOREWORD

ADVANCES IN CHEMISTRY SERIES was founded in 1949 by the American Chemical Society as an outlet for symposia and collections of data in special areas of topical interest that could not be accommodated in the Society's journals. It provides a medium for symposia that would otherwise be fragmented, their papers distributed among several journals or not published at all. Papers are refereed critically according to ACS editorial standards and receive the careful attention and processing characteristic of ACS publications. Papers published in ADVANCES IN CHEMISTRY SERIES are original contributions not published elsewhere in whole or major part and include reports of research as well as reviews since symposia may embrace both types of presentation.

CONTENTS

Preface viii

1. Devolatilization of Coal by Rapid Heating 1
   M. Menster, H. J. O'Donnell, S. Ergun, and R. A. Friedel

2. Production of Low Btu Gas Involving Coal Pyrolysis and Gasification ... 9
   C. Y. Wen, R. C. Bailie, C. Y. Lin, and W. S. O'Brien

3. Deuterium and Carbon-13 Tagging Studies of the Plasma Pyrolysis of Coal 29
   V. J. Krukonis, R. E. Gannon, and M. Modell

4. Arc Synthesis of Hydrocarbons 42
   Charles Sheer and Samuel Korman

5. The Reaction of Atomic Hydrogen with Carbon 54
   Alan Snelson

6. Problems in Pulverized Coal and Char Combustion 72
   David Gray, John G. Cogoli, and Robert H. Essenhigh

7. Coal Devolatilization in a Low Pressure, Low Residence Time Entrained Flow Reactor 92
   R. L. Coates, C. L. Chen, and B. J. Pope

8. Pressurized Hydrogasification of Raw Coal in a Dilute-Phase Reactor ... 108
   Herman F. Feldman, Joseph A. Mima, and Paul M. Yavorsky

9. Chemistry and Physics of Entrained Coal Gasification 126
   R. L. Zahradnik and R. J. Grace

10. Kinetics of Bituminous Coal Char Gasification with Gases Containing Steam and Hydrogen 145
    J. L. Johnson

11. Catalysis of Coal Gasification at Elevated Pressure 179
    W. P. Haynes, S. J. Gasior, and A. J. Forney

12. Alkali Carbonate and Nickel Catalysis of Coal–Steam Gasification ... 203
    W. G. Willson, L. J. Sealock, Jr., F. C. Hoodmaker, R. W. Hoffman, D. L. Stinson, and J. L. Cox

13. Catalyzed Hydrogasification of Coal Chars 217
    N. Gardner, E. Samuels, and K. Wilks

14. Thermal Hydrogasification of Aromatic Compounds 237
    P. S. Virk, L. E. Chambers, and H. N. Woebcke

Index 259

PREFACE

Until recently, interest in coal and other hydrocarbon gasification in the United States was limited mostly to a few technically trained people in a few research departments and institutions, and to the U. S. Department of the Interior's Bureau of Mines and Office of Coal Research. Projections of serious shortages of natural gas and other forms of energy during the 1950's and 1960's failed to spawn a positive response from the public and business communities. Even at this writing, in the face of severe fuel oil shortages and with the threat of gasoline rationing, many people believe the "energy crisis" to be fabricated and controlled by oil and natural gas interests.

Although the Arab oil embargo of 1973 sharpened the focus of public attention on the general energy situation of the United States, the energy problem was destined to be recognized. The United States is running out of reserves of oil and gas while it has enormous reserves of coal and oil shale. Undoubtedly we must turn to these solid fossil fuels as a major source of non-polluting energy while developing geothermal, solar, and nuclear forms of energy.

Evidence has emerged in the last two years to show an acute awareness of the need for clean gaseous fuels from coal on the part of the gas, petroleum, and electric power industries and on the part of the governments. A good example is the $120 million process development program sponsored jointly by the American Gas Association and the U.S. Department of the Interior's Office of Coal Research. Its goal is to have adequate information by 1976 to design and construct a coal gasification demonstration plant. Other large scale efforts are being launched by private companies and by new U.S. Government proposals for energy research, development, and, ultimately, demonstration plants.

Such extensive change in public and private attitudes toward energy and its efficient use is reflected in the scientific and technological communities where a much expanded and intense interest has developed rapidly. This is easily observed and, to a certain extent measured, by strongly increasing attendance at coal gasification symposia and the increasing volume of pertinent technical literature.

The need for fundamental information such as reaction mechanisms, catalysis, and kinetics and for new approaches such as the plasma arc for gasification studies has been clearly established. In partial satisfaction of this need, this volume presents 14 papers by distinguished authors, all aimed at providing fundamental scientific information of permanent value in the arduous search for substitute natural gas from coal.

L. G. MASSEY

Cleveland, Ohio
December 1973

1

Devolatilization of Coal by Rapid Heating

M. MENTSER, H. J. O'DONNELL, S. ERGUN, and R. A. FRIEDEL

Pittsburgh Energy Research Center, Bureau of Mines, U. S. Department of the Interior, 4800 Forbes Ave., Pittsburgh, Pa. 15213

Coals were devolatilized at rates comparable with those encountered in combustion and gasification processes. Rapid pyrolysis was attained with pulse-heating equipment developed for this purpose. This technique permitted control of the heating time and the final temperature of the coal samples. Subbituminous A to low volatile bituminous coals were studied. All bituminous coals exhibited devolatilization curves which were characteristically similar, but devolatilization curves of subbituminous A coal differed markedly. The products of devolatilization were gases, condensable material or tar, and residual char. Mass spectrometric analysis showed the gas to consist principally of H_2, CH_4, and CO. Higher hydrocarbons, up to C_6, were present in small quantities.

Coal has been used mainly for the generation of electric power. Now that there are critical shortages of natural gas for residential heating and industrial use, industry and government are developing processes for gasifying coal (1, 2). When these processes are fully developed, they will represent a second major outlet for coal utilization.

In both combustion and gasification, coal is heated to elevated temperatures therefore sustaining some degree of decomposition prior to or concurrently with other chemical reactions. In the Synthane process of the Bureau of Mines (3, 4), for example, pretreated coal enters the upper, carbonizing section of the gasifier where it undergoes extensive thermal degradation to form char. The reaction products formed at this stage in the process make an important contribution to the overall performance of the gasifier. For these reasons, research on the devolatilization of coal by rapid heating has been a part of our program on gasification. The results presented in this paper represent a continuation of preliminary work that was reported earlier (5).

Experimental

Single pulses of electrical current provide high-speed heating that is needed to measure the thermophysical properties of solids (metals in particular) (6, 7) at elevated temperatures (8). This technique, termed pulse heating, was adopted for devolatilizing coal in the present study. Coal samples were decomposed *in vacuo* in order to collect and identify the gaseous products. Quantitative measurements of the resultant weight loss of sample after rapid heating served as a measure of the total volatiles evolved from the coal.

The reaction vessel was essentially a 29/42-tapered, ground-glass joint sealed to a pumping system. Suitable vacuum gauges, manometers, and gas sampling and storage bulbs were attached to the reactor. Total volume of the reactor, including the sampling bulb, was 418 cm^3. No. 10 copper wires entered the reactor through Kovar-borosilicate glass seals. These copper electrodes terminated in spring clamps which supported the heating element containing the coal sample. Resistive heating elements were made into long, thin cylinders by wrapping 400-mesh stainless steel screen on a mandril. The cylinders were 6 cm long and 1.2 mm in diameter. In preparation for pulsing, the open ends of the cylinder were completely closed, and the flattened ends were inserted into the jaws of the spring clamps.

Current was supplied to the wire-screen heating elements by a current controller. This device was an electronic circuit designed to set the initial current flow at a desired value and to allow the current to increase in a predetermined way. Shaping of the current pulse was necessary to compensate for increase in electrical resistance of the wire and also for radiant heat losses at high temperatures. Typical current values were in the range 15–20 amps. The current controller was triggered by a preselected pulse coming from a General Radio unit-pulse generator, and current flow continued only for the duration of the timing pulse. Pulse times extended from 65 to 155 msec. A 0.1 Ω resistor in the current controller converted the current pulse to a voltage pulse which was displayed on a storage-type oscilloscope. Precise values of current and time were measured from the oscilloscope trace.

Coal samples were prepared by cutting vitrains from lumps of coal. The vitrains were further upgraded by microscopic examination in which coal particles with adhering mineral matter were discarded. Vitrains were chosen for study because they constitute the most abundant and homogeneous component of coal and because they are also low in min-

Table I. Proximate Analyses of Vitrains

		Proximate Analysis, % (mf)		
Coal Source	Rank	Fixed Carbon	Volatile Matter	Ash
Pocahontas No. 3, W. Va.	lvb	82.4	16.8	0.8
Lower Kittanning, Pa.	mvb	73.8	25.3	0.9
Pittsburgh, Pa.	hvAb	63.1	35.1	1.8
Colchester Illinois No. 2, Ill.	hvCb	51.1	48.0	0.9
Rock Springs No. 7½, Wyo.	Sub A	61.7	37.7	0.6

eral matter (9). The ash content was less than 2% in the Pittsburgh coal and less than 1% in the other coals used (Table I). A low mineral-matter content in the vitrains was desired to avoid ambiguities in the data from possible pyrolysis of mineral matter. The vitrains were ground to 44–53 μm particle size for the experiments.

Coals were selected to encompass a range of rank and volatile matter. Bituminous coals ranged from hvCb to lvb; one subbituminous coal was also studied. Their proximate analyses are given in Table I.

The temperature attained by the wire-screen heating elements was related to the time of current flow by a calibration method. Times required to melt pure metal powders of like particle size and amounts as the coal were determined by trial and error. A number of calibration points were thus established, and the temperature at the end of the current pulse was proportional to the time of current flow in the region to 1450°C. The heating rate was therefore a constant 8250°C/sec.

A new (unheated) screen cylinder containing no coal was pulsed to 900°C in the reactor which had previously been evacuated. Prefiring of the screen cylinders is essential because they undergo significant weight losses when they are heated for the first time. Such losses would interfere with measurements made on the coals. However, after the initial heating of a screen cylinder, its weight remains demonstrably constant in further tests. A prefired screen was weighed precisely on a Cahn RG microbalance, approximately 250 μg of coal was inserted into the cylinder, and the combined weight of the screen cylinder and coal sample was again determined precisely on the balance. The weighed coal sample and heating element were placed in the reactor and pumped until the system pressure was reduced to 10^{-3} torr. When this reduced pressure was attained, the coal sample was pulse-heated to a given temperature. After devolatilization occurred, the coal residue and screen were removed from the reactor and reweighed.

The volume of gases generated during devolatilization was determined from the pressure increase in the reactor. Mass spectrometric analyses of the gases were made at many, but not all, of the different test conditions. In this way, the weight of the gases produced by rapid devolatilization of coal was ascertained.

Results and Discussion

The devolatilization behavior of bituminous coals under rapid heating conditions is shown in Figure 1. This figure presents the weight-loss curves of four bituminous coals of different rank over temperatures from 400° to 1150°C. All of the weight-loss curves have a characteristic shape in common, although they differ in detail. For most of the coals the reaction threshold occurs at 400°C, followed by very rapid decomposition to 600°C. Production of volatile reaction products reaches a peak at relatively low temperatures of 700°–900°C, a finding that should be of considerable importance to those engaged in design of coal-gasification equipment. At still higher temperatures the declining trend in the forma-

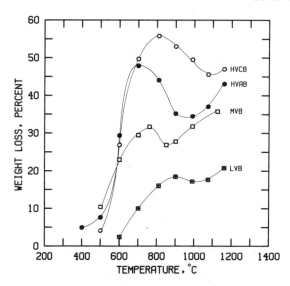

Figure 1. Devolatilization of bituminous coals by rapid heating

tion of volatiles reverses and starts to increase again at the highest temperatures of this study.

Some discussion of the low temperature peak in volatile production from bituminous coals is merited because this phenomenon does not occur during slow heating. In the latter case the weight loss increases monotonically with temperature (10). The broadest peak is exhibited by the Illinois hvCb coal which has the highest volatile-matter content of the coals studied (see Table I). For the higher rank bituminous coals, the peaks become progressively less intense with increase in rank, and the peak position shifts to higher temperatures. In all instances the volatile yield corresponding to the peak in the weight-loss curves was greater than the volatile matter in the coal determined by the ASTM standard method of analysis. These results are demonstrated by the data in Table II which show that the yield of total volatiles may be increased

Table II. Increased Volatiles from Rapid Pyrolysis

Coal Source	Volatile Matter Content, %		Increase Factor
	By ASTM Analysis	From Peak Weight Loss	
Pocahontas No. 3	16.8	18.5	1.10
Lower Kittanning	25.3	30.8	1.22
Pittsburgh	35.1	47.9	1.36
Colchester Ill. No. 2	48.0	55.8	1.16
Rock Springs No. 7½	37.7	42.4 (plateau)	1.12

as much as 36% by rapid heating. Supportive evidence for increased yields of volatiles is found in other rapid heating studies (11, 12).

Consideration of the cited studies in conjunction with our own leads to the conclusion that the ratio of total volatiles from rapid heating to ASTM volatile content depends not only on the rank of coal, as shown in Table II, but also on the magnitude of the heating rate. One suggested explanation for the appearance of maxima in the weight-loss curves is that of competitive reactions. For example, the bond-breaking reactions that occur in the coal structure and give rise to initial decomposition fragments may well have different temperature dependencies from those of recombination reactions that may form molecules more stable than the parent coal.

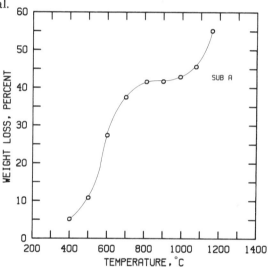

Figure 2. Devolatilization of subbituminous coal by rapid heating

In contrast to the results obtained with bituminous coals, the weight-loss curve of subbituminous coal exhibited no peak; instead, it reached a plateau in Figure 2. From 800° to 1000°C the volatile yield remained level at about 42 wt % of the coal. Beyond this region the production of volatiles increased sharply. The fact that the devolatilization curve of subbituminous A coal differs distinctly from those of bituminous coals indicates a need for further study of other subbituminous coals and lignites. Low rank materials such as these are of interest in coal gasification because their reserves are abundant and because they are situated in deposits with shallow ground cover.

The gases arising from rapid pyrolysis of coal vitrains have been examined by mass spectrometric analysis. The major components in the gas are H_2, CH_4, and CO. Lesser amounts of CO_2 and the higher molecular weight hydrocarbons (up to C_6) are also present. Hydrocarbons are present as both saturates and unsaturates with the notable exception of acetylene. Traces of aromatics such as benzene, toluene, and xylene are found as well as sulfur in the form of H_2S. The absence of acetylene, which has been found in appreciable quantity in some rapid heating processes (13, 14), is most likely attributable to the lower temperatures and lower heating rate employed in our experiments.

Temperature profiles of the individual gases from pyrolysis of Pittsburgh vitrain are shown in Figure 3. The molar percentages of CO_2, CH_4, and the C_2–C_4 hydrocarbons decrease with increasing reaction temperature. The functional dependence of H_2 and CO on temperature is more complex. H_2 production starts at 31.5 mole % at 700°C and increases to a maximum of 67.0 mole % at 990°C. Further increase in temperature causes a small but real decrease in its concentration. CO concentration changes in an opposite manner to H_2. A minimum CO value of 12.0 mole % is achieved at about the same temperature at which the maximum H_2 concentration occurred. The gas composition data are given on a H_2O–O_2–N_2-free basis.

In addition to the gases produced by rapid devolatilization of coal, heavier products, referred to as tar, also form. This material condenses on the walls of the reactor and is visible as a brown stain on the glass.

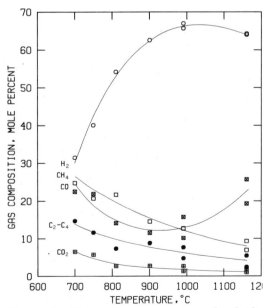

Figure 3. Composition of gas from devolatilization of Pittsburgh hvAb coal

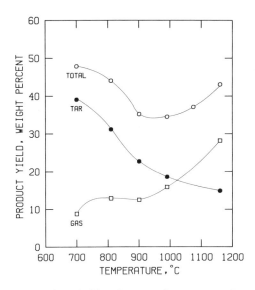

Figure 4. Yields of tar and gas from devolatilization of Pittsburgh hvAb coal

Because the quantity of tar from a single experiment is so small, it has not been measured nor has it been chemically analyzed. However, the quantity of tar can be obtained indirectly by subtracting the weight of the gases from the total volatiles, *i.e.*, the weight loss of the coal. Results of such calculations for Pittsburgh vitrain are shown in Figure 4 in which the experimentally determined curves for total volatiles and for gas have first been drawn. The curve for tar is of course determined by the difference calculation. The curves show that tar formation is favored by low decomposition temperatures and that tar is in fact the main product at all temperatures up to 1000°C. Above 1000°C the amount of gas exceeds the amount of tar even though the total volatile yield is still below the peak yield obtained at 700°C. Further correlations of product yields with rank and temperature parameters have been made and will be published later.

Literature Cited

1. Forney, A. J., Haynes, W. P., *Amer. Soc. Mech. Engrs., Meetg.*, New York, Nov. 26-30, 1972, paper **72-WA/APC-3**.
2. Henry, J. P., Louks, B. M., *Chem. Tech.* (1971) **1**, 238.
3. Forney, A. J., Gasior, S. J., Haynes, W. P., Kattel, S., *U. S. Bur. Mines Tech. Prog. Rept.* (1970) **24**, 6 pp.
4. Forney, A. J., Haynes, W. P., Elliott, J. J., Kenny, R. F., *U. S. Bur. Mines Tech. Prog. Rept.* (1972) **48**, 6 pp.
5. Mentser, M., O'Donnell, H. J., Ergun, S., *Amer. Chem. Soc., Div. Fuel Chem., Prepr.* **14** (5), 94 (Sept. 1970).

6. Cezairliyan, A., Morse, M. S., Berman, H. A., Beckett, C. W., *J. Res. Nat. Bur. Stds.—A. Phys. Chem.* (1970) **74A**, 65.
7. Cezairliyan, A., *J. Res. Nat. Bur. Stds.—C. Eng. Instr.* (1971) **75C**, 7.
8. Finch, R. A., Taylor, R. E., *Rev. Sci. Instr.* (1969) **40**, 1195.
9. Parks, B. C., O'Donnell, H. J., *U. S. Bur. Mines Bull.* (1956) **550**, 25.
10. Van Krevelen, D. W., "Coal," Elsevier, Amsterdam, 1961, p. 266.
11. Kimber, G. M., Gray, M. D., *Combust. Flame* (1967) **11**, 360.
12. Field, M. A., Gill, D. W., Morgan, B. B., Hawksley, P. G. W., *BCURA Mon. Bull.* (1967) **31**, 193.
13. Karn, F. S., Friedel, R. A., Sharkey, A. G., Jr., *Fuel* (1969) **48**, 297.
14. Fu, Y. C., Blaustein, B. D., *Ind. Eng. Chem., Process Des. Devel.* (1969) **8**, 257.

RECEIVED May 25, 1973.

2

Production of Low Btu Gas Involving Coal Pyrolysis and Gasification

C. Y. WEN, R. C. BAILIE, C. Y. LIN, and W. S. O'BRIEN

Chemical Engineering Department, West Virginia University, Morgantown, W. Va. 26506

Experiments involving the pyrolysis of bituminous coal, sawdust, and other carbonaceous feed materials have been performed in a 15-inch diameter, atmospheric, fluidized bed. Data from the pyrolysis experiments are analyzed to generate kinetic and heat-transfer information and to formulate a coal pyrolysis model useful in the design of commercial-sized processes. The model is then applied in forming a conceptual flowscheme for a relatively low pressure (5-13 atm) electrical-power generation plant. In the conceptual flowscheme, the low Btu gas is produced in two units, a pyrolyzer and a pyrolysis-char gasifier. The gas is then purified and fed into a combustion chamber; the electricity is generated in an advanced design gas turbine and steam turbine power cycle.

The demand for electrical power in the United States is predicted to quadruple in the next 20 years, with fossil fuels expected to be the energy source for at least half of this fourfold increase. Coal, because it represents over 95% of the untapped fossil fuel reserves, will certainly serve as the primary domestic energy source for most of these additional power requirements. However, uncontrolled coal burning is a dirty process with solid flyash particulates, sulfur dioxide, and nitrogen oxides as the major pollution culprits (1). The combined efforts of industry and government agencies are urgently needed to develop economical, efficient, and environmentally acceptable methods to convert coal into clean electrical power.

In this chapter, we describe a scheme to pyrolyze caking coal in a fluidized bed, present experimental data, devise a pyrolysis-gasification reaction model, and offer a conceptual flowscheme to convert coal to electricity *via* the production of low Btu gas. Although the experimental data presented here are not comprehensive, we will discuss some of the process path alternatives in such a manner as to recognize the most efficient ways to maximize the coal utilization efficiency.

Experimental

Equipment. An experimental, 15-inch diameter fluidized bed was used at West Virginia University to study the pyrolysis of coal and other carbonaceous compounds. The scheme of the pilot-plant fluidized-bed reactor and its auxiliary equipment is shown in Figure 1.

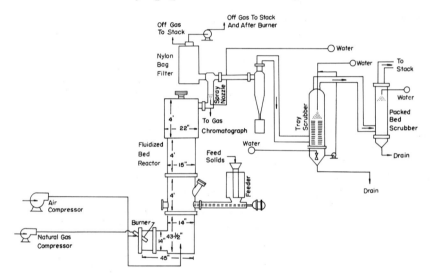

Figure 1. Fluidized-bed pyrolysis reactor system

There are three sections in the fluidized-bed reactor. In the L-shaped hot-bottom chamber, natural gas (more than 90% methane) is burned, and the hot combustion gases are mixed before passing through the grid plate to fluidize the sand bed. The composition of the combustion product gas can be adjusted, within a limited range, with additional air to form specific component ratios.

Between the gas-combustion chamber bottom and the reactor bed section is the high temperature grid plate. This gas-distribution plate is 1/4-inch thick, 18 inches in diameter, and made of Type 310 stainless steel. There are 584 holes in a square pattern located on 1/4-inch centers. Each hole is 0.0960 inches in diameter.

In the middle of the reactor is the fluidized sand bed. The sand serves two major purposes in the experiment. The violent agitation of the sand provides extremely rapid heat transfer to the carbonaceous fuel particles which undergo abrasion and grinding that constantly exposes fresh reactive solid surfaces as well as reduces the tendency for the reacting fuel particles to agglomerate. Also, the sand bed acts as a massive heat sink to dampen temperature fluctuations caused by unexpected upsets in the experimental systems.

This reactor section containing the fluidized bed is 28 inches od and 15 inches id. The insulation lining is $4\frac{1}{2}$ inches of Type 1620-K fire brick (Babcock and Wilcox Co.) and 2 inches of Plicast Tuff-Mix castable refractory (Plibrico Co.). The height of this section is 8 ft. Above the fluidized bed section, the reactor diameter expands to 35 inches od and 22 inches id. This particle-disengaging chamber is 4 ft high and is lined in the same manner as the fluidized bed section.

After leaving the reactor, the effluent gases are cooled and then cleaned by passage through either a canister-type nylon-bag filter or a dry-gas cyclone (10 inches in diameter and 22 inches long). Before being finally exhausted to the building exterior, the effluent gases are scrubbed in a series of two wet scrubbers: the first is a tray-type and the second is packed with 1-inch Intalox Saddles (U. S. Stoneware).

The fuel solids are fed into the fluidized bed by means of a screw conveyor with a specially designed feeder valve. The feed port is located 5 inches above the gas distribution plate. The $2\frac{1}{2}$-inch screw is constructed of carbon steel with a Type 310 stainless steel coating on the wearing surfaces. The construction details of this feeder and the auxiliary solid feed hopper system have been described by Burton and Bailie (2). The feeding unit has performed successfully in smoothly feeding 15-60 lbs of solids/hr into the fluidized bed.

The gases leaving the fluid-bed reactor are sampled every 5 min and analyzed by Bendix Chroma-Matic Model 618 Process Gas Chromatograph (Process Instruments Division, The Bendix Corp.). This unit quantitatively analyzes the gas for H_2, CO_2, CO, CH_4, and O_2/argon. The O_2/argon value is the additive sum of the oxygen and the argon compositions because the gas chromatographic peaks of both components are identical when using a Molecular Sieve column. Periodically, grab-samples of the effluent gases were withdrawn and analyzed by a Beckman GC-2A gas chromatograph and a Fisher Scientific Co. gas chromatograph for the gas components listed above, plus acetylene, ethylene, ethane, and nitrogen.

Table I. Operating Conditions for Fluidized-Bed Pyrolysis Reactor

Operating temperature	$1400 - 1900°F$
Operating pressure	$0 - 10$ psig
Collapsed bed height	2.5 ft
Expanded bed height	$3.5 - 4$ ft
Average particle size of sand	0.025 inches
Density of solid sand particle	100 lbs/ft
Superficial fluidizing gas velocity	1.5 ft/sec

Operating Procedure. The reactor is filled with 0.025-inch diameter sand to a collapsed bed height of 30 inches. The gas velocity through the bed is maintained at a level where a good fluidization of the sand is assured and then the bed is heated to the preselected temperature (1840°F) by the combustion of methane in the bottom section of the reactor. The operating conditions in the reactor are summarized in Table I.

After the bed reaches the desired temperature, the rate of air to the methane burner and the rate of inert gas flow are adjusted to give the oxygen level and gas flow rate specified in the experimental plan. The reactor system is then allowed to come to steady-state conditions as judged by a leveling of the effluent gas composition read on the continuously operating process gas chromatograph and by constant bed temperatures.

The test begins by slowly introducing the solid fuel feed into the bed *via* the screw feeder. The bed temperature immediately drops because of the sensible heat required to heat the solid to the reaction temperature plus the heat of pyrolysis. The solid feed rate is carefully adjusted so that the bed temperature does not drop below the desired 1400-1500°F range. The reaction system is allowed to come to a new steady-state condition with a constant solids feed rate, and the feed rate of the solids is determined by weight difference.

Table II. Composition of Solid Feed Material

Composition of Solid Feed Material	Bituminous Coal	Sawdust
Moisture, wt %, wet basis	3.42	2.62
Ultimate analysis, wt %, dry basis		
Carbon	73.30	47.20
Hydrogen	5.34	6.49
Oxygen	10.23	45.34
Sulfur	2.80	—
Nitrogen	0.70	—
Ash	7.57	0.97
Heating value, Btu/dry pound	13,097	8814
Particle size, Harmonic mean diameter, μ	504	603

Experimental Data. The results of the pyrolysis experiments involving sawdust and coal (bituminous, Pittsburgh seam, high volatile A) are reported here. The compositions of these two solid feed materials are listed in Table II. The experiments with sawdust are described here to compare the composition of the effluent pyrolysis gases with those of the product gases evolving from the pyrolytic reaction with the coal. Four tests with sawdust and one test with coal are reported here. In addition, several other types of carbonaceous solids were pyrolyzed in the fluidized sand bed, including municipal solid waste, chicken and cow manure, rubber, plastic, and sewage sludge. The results of these other experiments were described by Burton and Bailie (*2*).

As described previously in this chapter, in the course of the reaction test there were two periods of steady-state reactor operation, the first period just before the solid is introduced into the fluid bed, and the second period occurring during the constant-rate solids feeding. In all five tests, the reactor operations just prior to feeding the solids were identical, as listed in Table III.

After the solids were fed into the reactor and after the system again reached steady-state conditions, the effluent gas was analyzed. The re-

Table III. Operating Condition Values During Pyrolysis Experiments

Inlet flow rate of air: 35.26 scfm
Inlet flow rate of natural gas: 3.40 scfm
Reactor temperature prior to feeding solids: 1840°F
Dry composition of gas exiting reactor before solids feed is started:

$H_2 - 0.11\%$ $CO_2 - 10.17\%$
$O_2/argon - 1.18\%$ $CH_4 - 0.07\%$

The remainder of the gas is assumed to be N_2.

Steady-State Conditions During Solids Pyrolysis

Condition	Coal Test	Sawdust Tests			
		A	B	C	D
Operating time under steady-state conditions, min	155	86	75	70	577
Reactor temp. °F	1430	1430	1460	1450	1500
Solids feed rate, dry lb/min	0.336	0.368	0.122	0.682	0.342

Table IV. Per Cent Composition of Effluent Gas (Dry) During Pyrolysis Experiments

Gas	Coal Test	Sawdust Tests			
		A	B	C	D
Measured by Process Gas Chromatograph					
H_2	4.95	4.58	2.50	6.03	5.21
CO_2	11.29	12.18	12.11	12.24	11.47
$O_2/argon$	0.89	0.81	1.07	0.83	0.93
CH_4	1.79	2.24	0.32	3.31	1.85
CO	2.24	7.54	2.21	11.50	7.57
Measured by Research Gas Chromatographs					
C_2H_2	0.22	0.53	0.07	0.96	0.56
C_2H_4	NM^a	NM	NM	0.07	0.05
C_2H_6	0.10	0.11	0.04	0.16	0.06
N_2	77.4	73.5	80.8	66.7	73.5

[a] NM-not measured.

sultant effluent gas composition values for each of the five experiments are given in Table IV.

The composition values of CO_2, $O_2/argon$, CO, CH_4, and H_2 were averaged from the analysis readings of the process gas chromatograph, and the composition values of C_2H_2, C_2H_4, C_2H_6, and N_2 were averaged from the analyses by the research gas chromatographs of several grab-samples taken during the duration of the test. Using these experimentally measured gas analysis values, a mass balance was computed about the reactor system, using the nitrogen flow rate as the basis of calculation. The mass balances were quite good considering the 2-5% accuracy of the flow-measuring meters and analytical instruments. The gas produced from the coal or sawdust pyrolysis is considered to be the net gas flow rate value, after subtracting the volumetric flow rate of the effluent gases prior to feeding the solids from the flow rates of the gases leaving the reactor during the solids pyrolysis reaction. These computed pyrolysis gas production values for the five experimental runs are listed in Table V.

Table V. Computed Pyrolysis Gas Compositions and Production Rates

Pyrolysis Gas Composition, vol %, dry	Coal Test	Sawdust Tests			
		A	B	C	D
H_2	46.9	25.6	37.5	23.6	30.0
CO_2	11.7	15.0	24.3	14.1	11.1
CH_4	16.6	12.4	3.72	11.9	10.5
CO	21.7	43.3	33.8	45.7	44.5
C_2H_2	2.08	3.05	1.04	3.82	3.28
C_2H_4	NM[a]	NM	NM	0.29	0.28
C_2H_6	1.01	0.65	0.54	0.63	0.32
Production Rate, scf/lb dry feed	10.9	18.3	18.2	16.0	18.6
Gas Heating Value,[b] Btu/scf	435	398	286	412	399

[a] NM-not measured.
[b] Net heating value excluding hot carrier gas.

The results of these experiments indicate that the pyrolysis of the coal will yield 10.9 scf of a 435 Btu/scf gas/lb of dry feed. This pyrolysis gas originates from the volatile portion of the coal particle, thus leaving the remaining carbon (and corresponding portion of the total heat content) in the solid form which can then be separated from the gas as particulate matter or from the bed as char and later be converted into synthesis gas in a secondary high temperature reactor. Similarly, 1 lb of dry sawdust can be pyrolyzed into 18.3 scf of a 398 Btu/scf pyrolysis gas. Based on the combustion heat of product gases compared with the combustion heat of the entering feed and neglecting the sensible heat of the effluent gas, the coal pyrolysis gas contains only 36% of the combustion heat content of the feed coal while the pyrolysis of sawdust converts about 90% of the incoming energy into the gaseous form, leaving very little remaining solid char.

Aspects of a Pyrolysis Reaction Model

Generalized Criteria for a Coal Pyrolysis Model. When a coal particle is pyrolyzed, the following products are generally found: gases such as CO, H_2, CH_4, C_2H_2, C_2H_6, and CO_2, condensible liquid hydrocarbons such as benzene and toluene, aqueous compounds, and solid char. When designing a coal conversion plant, one may design the reactor system to maximize the production of the gaseous hydrocarbon, the liquid hydrocarbon, or the char products. The slot-type coke oven is deliberately designed to maximize the char production by allowing the volatile gases to evolve slowly from the solid phase without exterior gas purging, thereby prolonging the gas-solid contact time.

Upon heating, coal becomes softened and forms a metaplastic. Simultaneous devolatilizations of the carbonaceous matter in the interior of the particle push the bitumen to the surface. If the heating rate is rapid, this phenomenon is so violent that the particle literally bursts and develops into a new solid form with a much larger surface area per solid mass. If the heating rate is slow, the products during prolysis tend to repolymerize into large, more thermally stable molecules of solid matter that are retained in the interstices of the residual char particle. At high temperatures, the products of pyrolysis are lower in molecular weight than those produced at lower temperatures.

The maximization of the condensible hydrocarbon production is reached when the evolved volatile product is quenched or cooled rapidly after leaving the solid phase, allowing a minimum of time for the larger molecules to thermally decompose into the lower molecular weight gases. Conversely, the synthesis gas production is maximized if the volatile hydrocarbon products are held at a high temperature for a prolonged period. This exposure to high temperature will crack the tars and other condensible molecules to lower chain aliphatics—CH_4, C_2H_6, C_3H_8, etc. The pyrolysis reaction mechanism has been discussed by a number of investigators (2, 3, 4, 5, 6, 7). Squires (7) cites experimental data reported by Schroeder (8) in which coal, catalyzed with 1% molybdenum and in a hydrogen atmosphere at 800°C, yielded a 42.2% liquid hydrocarbon fraction after a 5-sec gas residence time, a 23% liquid fraction yield after a 10-sec gas residence time, and, after a 25-sec residence time of the gas, the liquid fraction yield was only 9.9%.

Although the liquid fraction was not collected in the experiments while feeding coal or sawdust, a liquid fraction and a char fraction were collected while pyrolyzing a municipal solid waste mixture. The liquid fraction represented 7.0% and the char fraction was 13.5% (moisture and ash-free weight basis) of the inlet solid feed. This contrasts with the data reported by Sanner et al. (9) who destructively distilled a municipal refuse in a retort constructed to simulate a coke oven process. They found that a 900°C, the liquid fraction from the refuse was about 47% and the char fraction was close to 9%. The equipment used by Sanner and co-workers allowed the effluent gases to be cooled immedi-

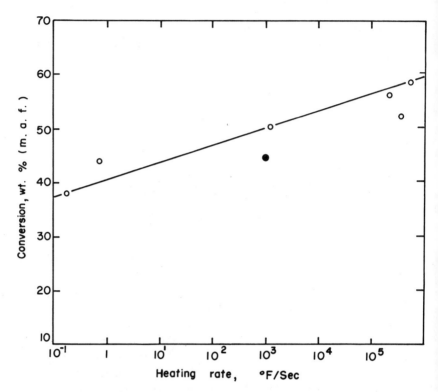

Figure 2. Effect of overall heating rate on devolatilization of coal
○ *Elkol coal* (5)
● *bituminous coal*

ately after leaving the retort while the exiting gases were held for more than 8-10 sec at close to the reaction temperature after leaving the pyrolysis zone of the fluid-bed reactor, thus accounting for the condensible hydrocarbon fraction from the fluid bed being a 40% smaller value.

In Figure 2 the final attainable conversion of coal as a function of overall heating rate is shown. This figure indicates that rapid heating of the coal avoids the polymerization reaction which can turn the coal into stabilized char before volatile matter is evolved. Figure 3 also reflects this observation. In Figure 3 the attainable conversion is shown as a function of the bed temperature. Here the conversion increases with increasing bed temperature. It is expected that higher bed temperatures will give high heating rates in the coal particles; hence the data of Figure 3 reinforce the observation made from the data presented in Figure 2.

A Simple Model of Coal Pyrolysis. In this section, a simple mathematical model of coal pyrolysis is formulated to qualitatively describe the pyrolysis of coal in a fluidized bed. This model is based on the assumptions that the devolatilization of a coal particle during the pyrolysis reaction is closely related to the heating rate of the particle and that the product distribution is primarily determined by the vapor residence time in the pyrolysis unit. As noted in the previous section, Figures 2 and 3 demonstrate this first point.

In view of the first assumption given above, it will be demonstrated that very rapid heating rates (hence, high fractional conversions of the coal to volatile matter) can be obtained in a fluidized-bed pyrolysis unit. A simplified model of coal pyrolysis may be formulated as follows:

A simplified energy balance for the coal particle is given by

$$\rho_0(1-X)C_{Ps}\frac{dT}{dt} = \frac{3h_c}{R}(T_b - T) + \frac{3\sigma Fe}{R}(T_b^4 - T^4) - \Delta H \rho_0 \frac{dX}{dt} \quad (1)$$

The rate at which coal is converted to vapor products is assumed to be proportional to the amount of the unconverted coal that can be even-

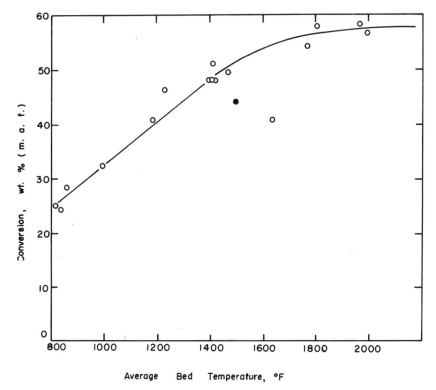

Figure 3. Effect of temperature on final conversion of coal attainable in fluidized-bed reaction

○ Elkol coal (5)
● bituminous coal

tually vaporized at set conditions. The kinetic reactions of the thermopyrolysis of cellulosic materials have been discussed rather thoroughly by Maa (10) who reported that several experimenters found the reaction to follow a first-order kinetic pathway for both wood (11, 12, 13, 14) and for other cellulosic materials (15, 16, 17, 18, 19). In a review of the reaction kinetics characteristics of a coal particle pyrolysis, Gray et al. (20) concluded that for particles smaller than 50 μm, the pyrolysis rate would be independent of particle size and first order with respect to unreacted volatile matter. This conclusion was based on previously reported experimental studies (21).

$$\frac{dX}{dt} = k_0 e^{-E/\tilde{R}T}(f-X) \tag{2}$$

In the derivation of Equation 1 it is assumed that the radius of the coal particle is so small that radial temperature variations within the coal particle may be neglected. Also the assumption of constant physical properties is made.

Appropriate initial conditions on the particle temperature, T, and the fractional conversion, X, are

$$T = T_0 \text{ and } X = 0 \text{ at } t = 0 \tag{3}$$

Equations 1 and 2, subject to the initial conditions given by Equation 3, can be solved numerically. However, an analytical expression for the particle temperature, which agrees very well with the numerical solution, can be obtained as follows:

The heat of pyrolysis, ΔH, is normally rather small, about 300 Btu/lb of coal. Thus, the heat generation term in Equation 1 can be ignored in comparison with the heat transferred to the particle by radiation and convection. Also, it is expected that the greatest heating rates will occur initially when the fractional conversion, X, is approximately zero. Hence, the variation in the particle density will be ignored in the heat accumulation term. This will lead to a conservative estimate of the heating rate. The above assumptions greatly simplify the solution of Equations 1 and 2 in that the equations become uncoupled. The equations can be further simplified by neglecting the heat transferred to the particle by convection in comparison with that transferred by radiation. This is usually a reasonable assumption in fluidized beds operating at high temperatures when small particles are injected into the bed.

Under the above assumptions, Equation 1 becomes

$$\frac{dT}{dt} = \frac{3e\sigma F}{R\rho_0 C_{PS}}(T_b^4 - T^4)$$

The solution of this equation is given by

$$1/2 \ln\left[\frac{(T_b+T)(T_b-T_0)}{(T_b-T)(T_b+T_0)}\right] + \tan^{-1}\left[\frac{T_b(T-T_0)}{T_b^2+TT_0}\right] = \frac{6Fe\sigma T_b^3 t}{\rho_0 C_{PS} R} \quad (4)$$

The conversion of coal to volatile material can be calculated by solving Equation 2 with the temperature given by Equation 4.

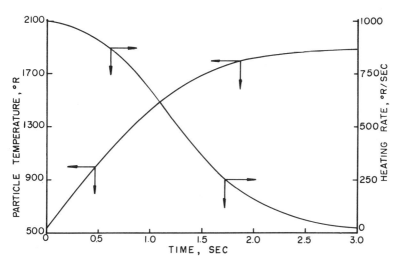

Figure 4. Calculated particle temperature and heating rate as function of time for pyrolysis of bituminous coal in fluidized bed

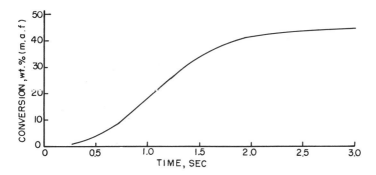

Figure 5. Calculated conversion of coal to volatile matter as function of time for pyrolysis of bituminous coal particle in fluidized bed

Figures 4 and 5 show the temperature of the particle and the per cent conversion of the coal particle to volatile material, respectively, as a function of time. The values of the parameters and the physical proper-

Table VI. Values of Parameters Used in the Calculation
of Figures 4 and 5—Bituminous Coal

Parameter	Value
ρ_0	50 lb_m/ft^3
C_{PS}	0.4 $Btu/(lb_m \cdot °F)$
R	0.000827 ft
k_0	$8.88 \times 10^4\ hr^{-1}$
E	8300 Btu/lb-mole
f	0.45
T_0	537°R
T_b	1900°R
$F \cdot e$	0.9

ties appearing in Equations 2 and 4 used in the calculation of these figures are given in Table VI.

Inspection of Figures 4 and 5 reveals that in a fluidized-bed pyrolysis unit, the heating rate of the particle is large and the particle temperature and conversion closely approach their final values within a few seconds. The high initial heating rate in the coal particle is believed to be the reason for the so-called creaming off of the hydrocarbons. With high heating rates, hydrocarbons can be driven off from the particle before they can undergo carbonization to form stable residual carbon.

Two Conceptual Coal-to-Low Btu Gas Conversion Flowschemes

A number of conceptual designs have already been proposed to convert coal to a low Btu gas and then use that gas to generate electrical power efficiently and cleanly. A modification of the high Btu Bi-gas process two-stage gasifier of Bituminous Coal Research has been proposed to use air instead of pure oxygen and to operate the gasifier at 300 psig. BCR concluded that an in-plant coal gasification process may compare favorably with other environmental control concepts (such as tail-end SO_2 removal) if the total coal-to-electricity process were to be redesigned into an optimal system (22).

In this section we intend to describe a conceptual process alternative based on the experimental data presented in the previous section, and to use this flowscheme to show that there will be a distinct advantage in considering a two-step coal gasification subsystem. In the first step, the coal is pyrolyzed to release the larger carbon molecules such as methane, ethane, and propane which are the cream of the decomposition products of the coal molecule. In the second-step gasifier vessel the residue pyrolysis char reacts with steam and air to form the gas containing H_2, CO, CO_2, etc., that is needed to fluidize the pyrolyzer.

The two processes compared here are illustrated in Figures 6 and 7. In the single-stage coal gasifier shown in Figure 6, the raw coal is fed

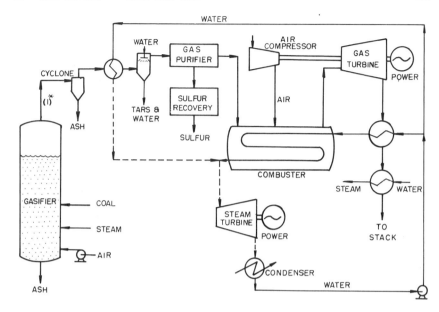

Figure 6. Power generation via coal gasification combined cycle (one-stage coal gasification system)

* Gas compositions are given in Table VII

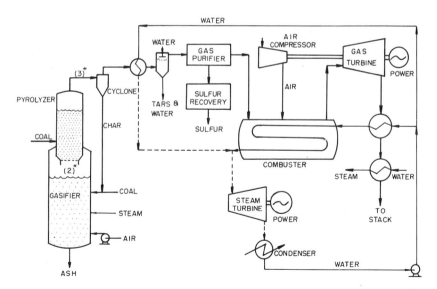

Figure 7. Power generation via coal gasification combined cycle (two-stage coal pyrolysis gasification system)

* Gas compositions are given in Table VII

directly into the high temperature (1900°F), fluidized-bed synthesis gas generator. The coal would have the same composition as that used in the pyrolysis experiments described earlier in this chapter (Table II). In the gasifier, the coal is directly gasified with air and steam to produce a stream of H_2, CO, H_2O, CO_2, CH_4, H_2S, and N_2. This product gas is cleaned of the H_2S and other impurities and is then burned in the combustion chamber. The effluent gases from the combustion chamber are sent through a combined gas turbine-steam turbine cycle. In this model, the gasifier system is assumed to operate adiabatically, the gases—H_2, CO, CO_2, H_2O—are assumed to emerge from the reactor in the same composition ratio as the equilibrium composition of the water-gas shift reaction, and the carbon-steam reaction gas components are assumed to vary from 4 to 6% in their approach to thermodynamic equilibrium.

The two-stage coal pyrolysis-gasifier is illustrated in Figure 7. The raw coal is blown into the fluidized bed where the pyrolysis of the coal takes place at 1400°F. It is conservatively assumed that there are no chemical reactions between the fluidizing gases and the volatilized coal-pyrolysis product char. It is also assumed that the product gases evolved from the coal pyrolysis reaction are produced at the same rate and in the same composition as those produced in the experimental bed described previously. The gas used to fluidize the pyrolysis unit contains hydrogen and is reducing in nature, so considerable hydrogasification should be expected. Therefore, the assumption that no reactions other than coal pyrolysis occur is certainly conservative. The char separated from the effluent gas then reacts with air and steam to produce the fluidizing gases for the coal pyrolyzer. The char-gasifier operates under the same conditions as described in the previous paragraph for the single-stage coal gasifier. A small amount of raw coal must be added to the char feed to the gasifier in order to maintain the 1900°F gasifier temperature and to produce enough gas to fluidize the coal particles in the pyrolyzer. The gas produced in this two-stage gasification system is then purified to remove the sulfur and other undesirable compounds and is burned in the combustion chamber with the combustion gases processed to generate electricity in the same gas- and steam-turbine system as described previously.

The compositions of the gas streams being produced in the three reactors of the two processes are listed in Table VII.

Table VII. Material Flow Rates and Gas Stream Compositions
Fuel: high volatile bituminous coal, analysis given in Table II
(A) Two-Stage Coal Pyrolysis-Gasification Process
  Overall thermal efficiency: 38.87%
  Proportion of electrical power generated: gas turbine —65.6%
                                              steam turbine—34.4%

Table VII. Continued

Flow Rates for 500 MW Power Plant

	lb-moles/hr	tons/hr
Coal feed to pyrolyzer	—	170
Coal feed to gasifier	—	3.7
Steam (150 psig) to gasifier	4,600	41.5
Air to gasifier	24,400	352
Char-ash for disposal	—	31.1
Gas: from gasifier	35,300	449
Gas: from pyrolyzer	44,800	518
Pyrolyzer char	—	101
Elemental sulfur sludge (50% water)	—	9.4
Gas: purified low-Btu gas	44,500	513
Air to combustion chamber	222,000	3202
Gas: effluent from combustion	256,300	3716
Water-steam: steam turbine cycle	51,000	459

Gas Stream Concentrations, mole %

Gas	Effluent from Gasifier	Effluent from Pyrolyzer	Gas from Purifier	Effluent from Combustion Chamber
CO_2	3.94	4.71	4.74	6.47
CO	30.09	27.15	27.33	—
H_2	11.34	18.55	18.67	—
H_2O	2.80	3.60	3.62	5.46
N_2	51.60	40.92	41.18	75.62
O_2	—	—	—	12.45
CH_4	0.21	3.74	3.77	—
H_2S	0.02	0.66	—	—
C_2H_2	—	0.45	0.46	—
C_2H_6	—	0.22	0.23	—
	100.00	100.00	100.00	100.00

(B) Single-Stage Coal Gasification Process
Overall thermal efficiency: 36.93%
Proportion of electrical power generated: gas turbine —60.7%
steam turbine—39.3%

Flow Rates for 500 MW Power Plant

	lb-moles/hr	tons/hr
Coal feed to gasifier	—	183
Steam (150 psig) to gasifier	3,700	33.4
Air to gasifier	41,400	597
Char-ash for disposal	—	32.8
Gas: from gasifier	64,700	780
Elemental sulfur sludge (50% water)	—	9.8
Gas: purified low-Btu gas	64,400	775
Air to combustion chamber	185,600	2676
Gas: effluent from combustion chamber	237,100	3451
Water-steam: steam turbine cycle	44,900	404

Table VII. Continued

Gas Stream Concentration, mole %

Gas	Effluent from Gasifier	Gas from Purifier	Effluent from Combust. Chamber
CO_2	4.08	4.10	7.73
CO	24.05	24.17	—
H_2	15.53	15.61	—
H_2O	4.97	4.99	5.70
N_2	50.69	50.93	75.64
O_2	—	—	10.93
CH_4	0.20	0.20	—
H_2S	0.48	—	—
	100.00	100.00	100.00

Effect on Overall Efficiency of Certain Process Variables

The combined power cycle scheme used in this study involves a gas turbine following the combustion chamber to expand the gas from the operating pressure (5-10 atm) to atmospheric pressure and then a steam cycle which utilizes the sensible heat of the gas turbine effluent and additional heat from the combustion chamber to superheat the steam. The total net work produced will vary with the amount of excess air provided to the combustion chamber. The maximum amount of net work is produced when the combustion air rate is enough to hold the combustion chamber at the predetermined operating temperature with little indirect cooling. The efficiency of the overall coal conversion process is defined as the net electrical power (Btu equivalent) produced, divided by the combustion heat (high heating value) of the fuel fed into the process as the raw energy source, in this case, coal.

As illustrated in Figure 8, the two-stage pyrolyzer-gasifier system generates electricity at a 1.9% better efficiency than does the single-stage gasifier system. This comparison was made with the synthesis fuel gas being cooled to relatively low temperatures (200-300°F) in the purification subsystem and reheated in the combustion chamber and with the gases entering the gas turbine at 2000°F. The maximum efficiency was found to be 38.87% for the pyrolyzer-gasifier process and 36.93% for the single-stage gasifier process. A major reason for the better efficiency in the pyrolyzer-gasifier system is attributed to the relative volume of gases processed through the units. The steam and other utility requirements of some subsystems are significantly dependent on the total gas flow amounts rather than on component concentrations. This is especially true with the steam requirements of the gas purification subsystem. The conceptual systems described above represent the operation of the gas purification subsystem with a commercially available sulfur-removal process which requires the pyrolysis gas stream to be cooled to relatively low temperatures (200-250°F) before being purified. A major effort has been made to recover, as generated steam, as much of the evolved heat energy as possible and to re-use this heat elsewhere in the process.

The ideal gas purification subsystem would operate at the temperature of the effluent gases from the coal-gasification units so that the cleaned gases can be blown hot into the combustion chamber with little loss in sensible heat. Several such hot-gas purification-cleaning processes have been recently proposed, such as using dolomite as the solids in the fluidized bed to adsorb the sulfur pollutants (Exxon Research) or operating a scrubbing stage with molten calcium salts (Battelle Northwest). If such a perfect system were developed to clean and purify the gases at high temperatures with negligible steam and power requirements, the efficiency improvements realized would be quite substantial, as shown in Table VIII.

Table VIII. Effect of Gas Purification Conditions on Overall Plant Efficiency

Gas Purification	Two-Stage Pyrolyzer-Gasifier, %	Single-Stage Gasifier, %
Low-temp	38.87	36.93
High-temp	40.77	39.51

The electricity-generating ability of the present-day gas turbine is limited by the temperature of the inlet gases. The maximum allowable operating temperature today is in the 1900-2000°F range and is governed by the thermal tolerance of the turbine construction metal. The values shown in Table IX indicate the effect that changes in this gas temperature will have on the overall process efficiency of the coal conversion processes. Figure 8 also illustrates this effect.

Table IX. Effect of Gas Turbine Inlet Temperature on Maximum Overall Plant Efficiency

Temp. of Gases Entering Gas Turbine, °F	Two-Stage Pyrolyzer-Gasifier, %	Single-Stage Gasifier, %
1800	36.6	35.0
2000	38.9	36.9
2200	40.6	38.5

Another very important design consideration is the degree of carbon utilization realized in the synthesis gas generator of both processes. The illustrated processes were designed utilizing 85% of the carbon in the

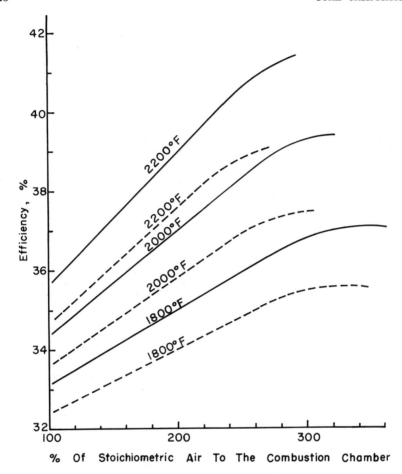

Figure 8. *Efficiencies for power generation via low Btu coal gasification processes and combined gas-steam cycle system one-stage process (gasification)*

——— *two-stage process (pyrolysis-gasification)*
- - - - *one-stage process (gasification)*

raw coal fed to the plant. An increase to 90% carbon utilization will mean an increase of approximately 1.6 ∼ 1.9% in the thermal efficiency of the overall coal-to-electricity conversion plant.

Conclusion

Experimental data of coal pyrolysis in a sand fluidized-bed reactor indicate that it is possible to extract considerable amounts of hydrocarbons from caking coal by a rapid heating and subsequent cracking in the vapor phase. This is done by flowing pulverized coal into a fluidized bed of hot sand and elutriating the char along with gaseous product from the fluidized bed. The extremely good heat transfer of the fluidized bed provides the rapid heating required for this process. In this manner, the valuable hydrogen in coal is extracted either as free hydrogen or as hydrocarbons in the gas-phase product. A conceptual scheme is presented which utilizes the product char to generate synthesis gas by gasification with air and steam for use in the pyrolyzer. The thermal efficiency calculated based on the two-stage process with the combined gas and steam turbine cycle shows that this scheme is a promising way to produce clean power from coal.

Acknowledgment

We thank W. J. McMichael for his help in revising portions of this manuscript.

Definitions

C_{ps}	Heat capacity of solid	(Btu/lb-°F)
E	Activation energy	(But/mole)
e	Emissivity of the coal particle	—
F	View factor for coal particle in fluidized bed	—
f	Final attainable conversion	—
h_c	Convective heat transfer coefficient	(Btu/ft²-hr-°R)
k_o	Frequency factor	(hr⁻¹)
R	Radius of particle	(ft)
$\tilde{R}$	Gas constant	(Btu/mole-°R)
T	Particle temperature	(°R)
T_o	Initial particle temperature	(°R)
T_b	Bed temperature	(°R)
t	Solid residence time	(hr)
X	Fractional conversion of solid to vapor products	—
ρ_o	Initial solid density	(lb/ft³)
σ	Boltzmann's constant	(Btu/hr-ft²-°R⁴)
ΔH	Heat of pyrolysis	(Btu/lb)

Literature Cited

1. Jimeson, R. M., *Amer. Chem. Soc., Div. Fuel Chem., Prepr.* **17** (1), 88 (New York, August-September, 1972).
2. Burton, R. S., Bailie, R. C., *Combustion*, submitted for publication.
3. Peters, W., Bertling, H., *Amer. Chem. Soc., Div. Fuel Chem., Prepr.* **8** (3), 77 (Chicago, September, 1964).
4. Kertamus, N., Hill, G. R., *Amer. Chem. Soc., Div. Fuel Chem., Prepr.* **8** (3), 89 (Chicago, September, 1964).
5. Jones J. F., Schmid, M. R., Eddinger, R. T., *Chem. Eng. Progr.* (1964) **60** (6), 69.
6. Kirov, N. Y., Stephens, J. N., "Physical Aspects of Coal Carbonization," University of New South Wales, Sidney, Australia, 1967.

7. Squires, A. M., *Ann. Meetg., AIChE, 65th,* New York, 1972.
8. Schroeder, W. C., U. S. Patent **3,030,297.**
9. Sanner, W. S., Ortuglio, C., Walters, J. G., Wolfson, D. E., Report of Investigations No. **7428,** U. S. Bureau of Mines, 1970.
10. Maa, P. S., Ph.D. Thesis, West Virginia University, Morgantown, W. Va., 1971.
11. Bamford, C. H., Crank, J., Malan, D. H., *Proc. Cambridge Phil. Soc.* (1946) **42,** 166.
12. Bowes, P. C., Fire Research Note No. **266,** Department of Scientific and Industrial Search and Fire Offices' Commission, Joint Fire Research Organization, Great Britain, 1956.
13. Stamm, A. J., "Wood and Cellulose Science," Ronald Press, New York, 1964; *Ind. Eng. Chem.* (1956) **48,** 413.
14. Wright, R. H., Hayward, A. M., *Can. J. Technol.* (1951) **29,** 502.
15. Kujirai, T., Akahira, T., *Sci. Papers Inst. Phys. Chem. Res. (Tokyo)* (1925) **2,** 223.
16. Madersky, S. L., Hart, V. E., Straus, S., *J. Res., Nat. Bur. Stds.* (1956) **56,** 343.
17. Madersky, S. L., Hart, V. E., Straus, S., *J. Res., Nat. Bur. Stds.* (1958) **60,** 343.
18. Martin, S., *Res. Devel. Tech. Rept.* **USNR DL-TR-102-NS081-001** (1956).
19. Murphy, E. J., *Trans. Electrochem. Soc.* (1943) **83,** 161.
20. Gray, D., Cogoli, J. G., Essenhigh, R. H., *Amer. Chem. Soc., Div. Fuel Chem., Prepr.* **18** (1), 135 (Dallas, 1973).
21. Howard, J. B., Essenhigh, R. H., *Ind. Eng. Chem., Process Design Devel.* (1967) **6,** 76.
22. Bituminous Coal Research, Inc., "Economics of Generating Clean Fuel Gas from Coal Using an Air-Blown Two-Stage Gasifier," R & D Report No. 20, Final Report—Supplement No. 1, U. S. Office of Coal Research, 1971.

RECEIVED May 25, 1973. Work supported partly by the Office of Coal Research, Department of the Interior and partly by the Solid Waste Office of Research and Monitoring, Environmental Protection Agency.

3

Deuterium and Carbon-13 Tagging Studies of the Plasma Pyrolysis of Coal

V. J. KRUKONIS and R. E. GANNON

Avco Corp. S/D, Lowell Industrial Park, Lowell, Mass. 01851

M. MODELL

Massachusetts Institute of Technology, Department of Chemical Engineering, Cambridge, Mass. 02139

In the conversion of coal to acetylene by the arc-coal process high product yields are obtained by quenching the high temperature gas mixture by a stream of hydrogen. Investigation of the role of the hydrogen quench, through the use of deuterium and ^{13}C additions, showed that the acetylene undergoes complete exchange of its atoms with other acetylene molecules as well as with the quench stream. A chain-reaction mechanism is suggested to account for the complete interchange of atoms.

High temperature arc or plasma pyrolysis of coal produces acetylene as the principal hydrocarbon product (1–6). Furthermore, yields of acetylene in a hydrogen atmosphere are enhanced by a factor of three over those achieved in an argon atmosphere. Consistent with the experimental results, thermodynamic data show that acetylene is the only stable hydrocarbon molecule above 1500°C, and below about 1200°C its thermodynamic stability decreases rapidly (7, 8, 9); experimental evidence also supports that high temperature acetylene-containing hydrocarbon streams must be quenched rapidly in order to prevent decomposition to carbon black (10).

During a program to convert coal to acetylene carried out at Avco Systems Divisions Laboratories, a number of high temperature arc reaction concepts were tested. The initial conversion scheme used coal as the consumable anode of a dc arc, the process schematic of which is shown in Figure 1. The consumable-anode pyrolysis of coal has been described in detail in Ref. 5; briefly, crushed coal, typically 10-20 mesh, is fed into an electrical discharge sustained between a graphite cathode and the coal at a feed rate consistent with the surface pyrolysis rate. The rapid heating occurring at the surface pyrolyzes the coal, and the hydrocarbon products formed are quenched downstream of the arc zone by injecting a gas to preserve the acetylene produced in the discharge region. The solid residue consisting of char and any unreacted coal spills over the sides of the anode feed tube.

A schematic diagram of the experimental reactor showing the coal-feed tube, gas-quenching ports, and the product-sampling positions is given in Figure 2. The gas-sampling tubes were located at sequential positions downstream of the arc zone in order to determine if any acetylene decomposition were occurring in the gas stream. Simultaneous sampling at all three positions shown in Figure 2 produced identical results. (Although probing of the high temperature arc zone with a small diameter, water-cooled tube produced higher acetylene concentrations, indicating that some decomposition was occurring even before the gas reached the first sampling position, the yield and decomposition data that are reported subsequently were obtained from downstream sampling positions and thus are not confused with ultra-high quench rate ambiguities.)

Hydrocarbon products were analyzed on an F&M 700 chromatograph equipped with a Porapak Q column and a flame ionization detector. (Other gases such as CO, H_2S, COS, and CS_2 were detected on an F&M

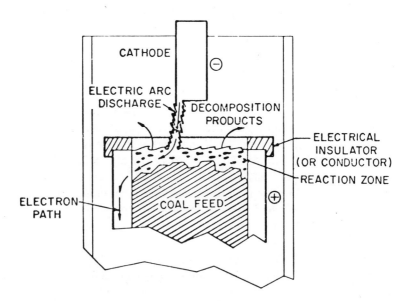

Figure 1.   Schematic of coal conversion arc process

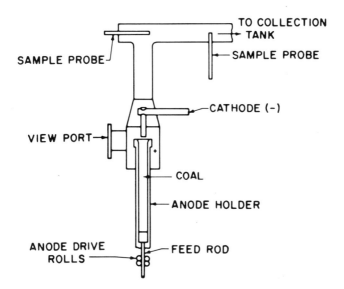

Figure 2. Arc-coal reactor

720 thermal conductivity chromatographic unit.) Data obtained using argon and hydrogen as quench gases are given in Figure 3 and show that at all power levels studied the yield of acetylene is 2-3 times greater with hydrogen as with the quench medium.

Pyrolysis of coal with consequent formation of acetylene in an inert plasma environment is evidence that the carbon and hydrogen present in the coal are reacting,

$$\text{C (in coal)} + \text{H (in coal)} \longrightarrow \text{C}_2\text{H}_2 \qquad (1)$$

although Equation 1 is obviously an oversimplification of acetylene formation, and no mechanism is implied.

Several explanations for the higher yields found with hydrogen can be proposed:

1. Hydrogen generates additional acetylene from the carbon in coal *via* solid carbon–gaseous hydrogen reaction, a yield contribution which is absent in an inert atmosphere.

$$\text{C (in coal)} + \text{H}_2 \longrightarrow \text{C}_2\text{H}_2 \qquad (2)$$

2. Hydrogen is a more effective coolant or preserver of acetylene than argon because of mass transfer or thermal conductivity considerations.

3. Hydrogen acts as some chemical reactant in the quenching step, preventing the decomposition of acetylene to carbon black.

The most obvious explanation for the improved acetylene yields found with a hydrogen quench is postulate 1—additional reactions occurring between carbon in the coal and the gaseous hydrogen environment—and this reaction contribution was subsequently tested by allowing char and hydrogen to react. Hydrogen-free char (collected from previous arc-coal tests) was allowed to react in the arc environment; however acetylene yields were only minimal (5), far below the levels required to explain the factor-of-three difference between the argon and hydrogen results shown in Figure 3.

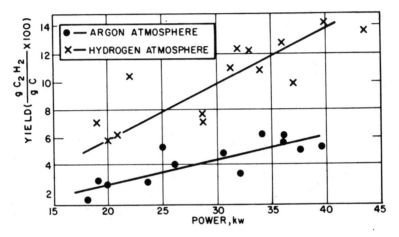

Figure 3. Effect of atmosphere on acetylene yield

Other experiments in the consumable-anode arc reactor showed that if no quench were used, acetylene yields were very small with substantial carbon black formation on the reactor walls; this indicated that decomposition of acetylene in the product stream was occurring, again in agreement with the thermodynamic data and other experimental evidence.

The coal pyrolysis scheme underwent a number of modifications in both reactor design and reaction philosophy during the course of the program. It was found, for example, that one of the most serious causes of acetylene decomposition (in either the argon or hydrogen case) was the contact between the incandescent surface char layer and the acetylene which was generated below the surface. Examination of large char particles showed that carbon black was present in the pores of the structure (6). Feeding hydrogen up through the coal increased the yields up to about 18% (based on total coal), but finally certain materials' erosion problems shifted emphasis from the consumable-anode concept to a plasma reactor shown in Figure 4. The reactor description and operation have been covered fully in Ref. 10; briefly, powdered coal, typically

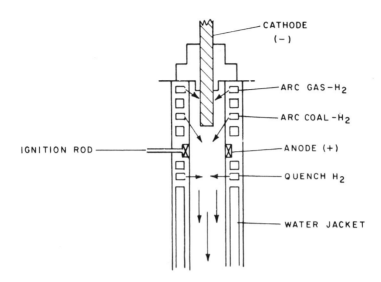

Figure 4. Rotating arc reactor

−100 mesh, is carried downward *via* hydrogen through a magnetically rotated arc region at a velocity of a several hundred ft/sec. A few inches downstream the hydrocarbon stream is quenched (and several alternatives employing coal, a hydrocarbon, or solely hydrogen were tested). Acetylene yields and electrical energy consumptions in this reactor approached economically viable levels; in the rotating arc reactor hydrogen still produced substantially higher acetylene yields than did an argon quench.

To separate the acetylene-forming step from the acetylene-preserving step in this coal pyrolysis scheme and also to determine quantitatively the role of hydrogen in the process, acetylene was injected into a carbon-free plasma stream, an experiment designed to simulate only the quenching-preserving step. In the quenching studies, as in the coal tests, the decomposition of acetylene in an argon plasma was again much greater than in hydrogen. Figure 5 gives the results of acetylene decomposition experiments which have been reported previously (10) and shows that about 60% of the initial acetylene decomposes with an argon quench while a hydrogen quench only about 10-12% of the acetylene decomposes.

Because of the large differences in results with hydrogen and argon, other quench gases—helium, nitrogen, and deuterium—were subsequently tested in an attempt to separate chemical effects from physical ones. For example, if gas diffusivity and thermal conductivity are important parameters in preserving acetylene as an intact species, acetylene decomposition in hydrogen, helium, and deuterium would be essentially the same because all three gases possess essentially identical mass and heat transport properties. However, if the heat capacity of the quench medium, *i.e.*, its energy absorption and dissociation capability, were the important consideration, hydrogen, deuterium, and nitrogen would produce identical quenching results (ignoring for now the somewhat higher stability of nitrogen relative to hydrogen). Finally, if chemical reactions between C_2H_2 and H_2 were in operation during the quenching step, H_2 and D_2 results would be identical. In addition to gas chromatography, high-resolution mass spectroscopy was carried out with the deuterium samples. The deuterium results which have been reported previously (*11*) showed that substantial H-D exchange was occurring in the acetylene molecule.

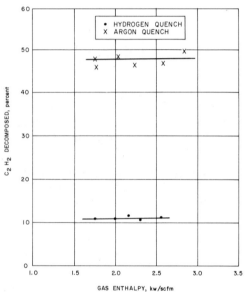

Figure 5. Decomposition of acetylene in quench simulator

Finally as a means of determining reactions in operation during the acetylene quenching step, heavy acetylene, $[^{13}H]C_2H_2$, and normal acetylene, $[^{12}H]C_2H_2$, were admixed and injected into the plasma quenching simulator so that carbon–carbon triple bond interaction at high temperature could be studied.

Results and Discussion

A series of experiments to separate the complex reactions of acetylene generation and acetylene preservation occurring during the arc pyrolysis

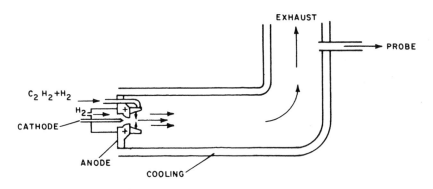

Figure 6. Quench simulator

of coal was carried out as an integral part of the coal pyrolysis studies in the rotating arc reactor. In order to study the quenching reaction, acetylene was injected into a high temperature plasma stream free of coal and char particles. A plasma generator attached to a 2-inch copper, water-cooled tube as shown in Figure 6 was used for the quenching studies. Acetylene was injected into the plasma stream about 1/2 inch from the nozzle, and a gas-sampling tube was located about 12 inches

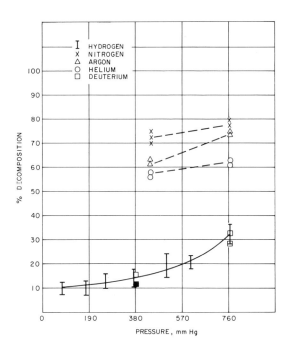

Figure 7. Decomposition of acetylene in various gases

downstream of the plasma generator; as in the previous reactors, the gas-sampling probe was purposely located far downstream of the plasma and injection position to allow all decomposition reactions to occur, eliminating the probing ambiguities described earlier. The acetylene concentration in the plasma was about 7%, closely approximating the concentration obtained in the rotating arc reactor. The entire plasma system was connected to a large capacity rotary vacuum pump so that tests could be performed at various total reaction pressures.

Tests were made then to determine the effect of operating pressure on decomposition and to determine the effects of using different gas quenches on preserving acetylene. Both sets of results are given in Figure 7. Most extensive testing was carried out with hydrogen, and decomposition is a function of total pressure but reaches only about 30% at 1 atm. Figure 7 also shows that decomposition of acetylene in both hydrogen and deuterium is identical at the two pressures studied and further shows that the decomposition is low relative to the other gases used; helium, argon, and nitrogen. The data in Figure 7 indicate then that hydrogen (and chemically similar deuterium) has some other effect in preserving acetylene, an effect not explainable by merely the physical properties of thermal conductivity, diffusivity, or heat capacity.

Table I. Isotopic Acetylene Composition

Species	Composition, % (Undecomposed C_2H_2)
C_2H_2	1.1
C_2HD	15.1
C_2D_2	83.5

There was no way to determine chromatographically if the acetylene molecules sampled in the hydrogen quenched stream were the identical molecules injected into the plasma; therefore the deuterium gas samples were analyzed by mass spectroscopy to determine if the composition consisted of C_2H_2 or some deuterated species. (Although it would have been possible, we did not attempt to chromatographically separate deuterated acetylenes.) The analyses of the deuterium samples showed that H-D interaction to form C_2HD and C_2D_2 was occurring during the plasma quenching step. Table I gives the measured composition of acetylene in the product gas for the deuterium plasma-deutrium quench set of tests.

Ninety-nine per cent of the C_2H_2 molecules have exchanged with D_2 to form C_2HD and C_2D_2.

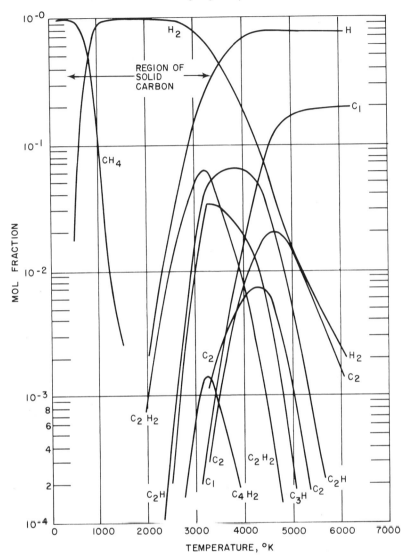

Figure 8. Equilibrium diagram for carbon–hydrogen system at 1 atm $(C/H = 1/4)$

A number of people studying very high temperature acetylene formation–decomposition have invoked radical recombinations to explain the fact that acetylene compositions greater than equilibrium were achieved in their studies. Plooster and Reed (12) postulated that high temperature equilibrium favors two carbon species, C_2H_2 and C_2H, and that during the quenching sequence two mechanisms contributed to the acetylene measured in the product.

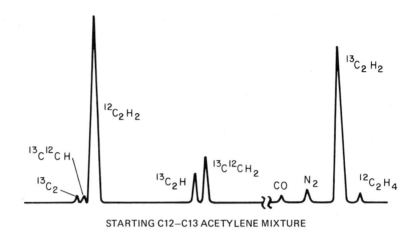

STARTING C12–C13 ACETYLENE MIXTURE

Figure 9. Mass spectra of

$$C_2H_2 \text{ (present at high temperature)} \xrightarrow{\text{quench}} C_2H_2 \quad (3)$$

$$C_2H + H \xrightarrow{\text{quench}} C_2H_2 \quad (4)$$

Their model was based on estimates of the thermodynamic properties, and their experimental results agreed well with the model of a C_2H radical. Later, Baddour and Iwasyk (13) and Baddour and Blanchet (14) also invoked the C_2H mechanism to explain their results of reacting hydrogen with carbon in a consumable anode arc reactor.

An equilibrium diagram for the carbon–hydrogen system is given in Figure 8 and shows that C_2H_2 and C_2H are in fact the most prevalent carbon species at high temperature. (Figure 8 was calculated for a C/H ratio of 1/4, i.e., for methane, using JANAF data (9).) Although calculated from free energy considerations to be present in relatively high concentrations, the C_2H radical has not yet been experimentally verified to be present in any quantity. A quenching mechanism which both rapidly cools C_2H_2 and requires a recombination of a nonintercon-vertical C_2H radical with H does not explain the presence of large amounts of C_2D_2 in the product stream. If C_2H maintains its identity at high temperature, only C_2HD (along with the preserved C_2H_2) could be formed during the quenching step; however, the data in Table I show that 83% of the C_2H_2 has exchanged to C_2D_2, and a random recombination of radical and atomic species was found to correlate the results.

Conservation of energy considerations, however, preclude the dissociation of all the species into atoms and/or radicals. For example, most of the tests in the plasma reactor were performed at input power levels

3. KRUKONIS ET AL. *Plasma Pyrolysis of Coal* 39

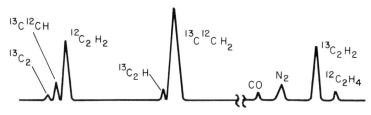

C12–C13 ACETYLENE MIXTURE AFTER EXPOSURE TO PLASMA

acetylene mixture

of about 15 kW; with a H_2 (or D_2) flow of 4.95 scfm used in the tests, the average gas enthalpy was only about 40 kcal/gram-mole, much less than the 100 kcal/gram-mole required to dissociate even hydrogen alone (not including the dissociation of acetylene to C_2 and H). In spite of the impossibility of dissociating all the molecules to C_2, H, and D species, however, there is presented in Table II the predicted statistical concentrations of H- and D-exchanged acetylenic molecules based on the recombination of such species, and the agreement is very good.

Table II. Isotopic Composition of Acetylene

Species	Measured Composition	Predicted Composition (Based on Combination of C_2, H, and D)
C_2H_2	1.1	0.5
C_2HD	15.1	13.2
C_2D_2	83.5	86.0

As a final study of the decomposition–recombination reactions of acetylene which could aid in the elucidation of the acetylene preservation mechanism in hydrogen, mixtures of carbon isotope acetylenes, [^{12}C and ^{13}C]C_2H_2, were injected into a hydrogen plasma stream, and analysis of the gas product again was determined by high-resolution mass spectroscopy. If the carbon–carbon triple bond maintained its integrity, only [^{12}C and ^{13}C]acetylene should be present in the product stream; if on the other hand the triple bond were entering into the decomposition–preservation reactions, an interaction to form [^{13}C, ^{12}C]C_2H_2 would be measured in the product.

For the [^{12}C–^{13}C] experiments the plasma reactor was also operated at about 0.5 atm with approximately a 50/50 mixture of the isotope acetylenes injected into the plasma stream. Gas chromatographic and high-resolution mass spectroscopic analyses were performed on the samples. Gas chromatographic analysis showed again that only about 10% of the original acetylene disappeared to carbon black (and the decomposition result is shown as a solid dot in Figure 7). Mass spectrograms of both the initial and final product acetylene streams are given in Figure 9. Figure 9 shows that some [^{12}C, ^{13}C]C$_2$H$_2$ was present in the starting material (because the heavy acetylene could be obtained only as 90% [^{13}C]acetylene). A mass spectrogram of the final composition obtained with the plasma generator operating at 15 kW, a power level identical to that of the deuterium tests, is also compared with the initial spectrogram in the figure, and a large increase in the [^{12}C, ^{13}C]C$_2$H$_2$ peak is evident. Acetylene compositions are given in Table III.

Table III. Isotopic Composition of Acetylene

Species	Initial Mixture (Power Off)	Final Mixture (Quenched)
	Composition %	
[^{12}C]C$_2$H$_2$	47.6	30.6
[^{12}C,^{13}C]C$_2$H$_2$	11.9	48.5
[^{13}C]C$_2$H$_2$	40.5	20.9

If 100% of ^{12}C and ^{13}C reacted, the calculated [^{12}C, ^{13}C]C$_2$H$_2$ concentration would be 49.6% as shown in Table IV. The measured concentration of 48.5% indicates therefore that a 97% exchange occurred based on the statistical model. Again, however, energy considerations (with the plasma stream possessing only a maximum of 40 kcal/gram-mole) preclude such a simple model.

Table IV. Isotopic Acetylene Composition

Species	Measured, %	Calculated, %
[^{12}C]C$_2$H$_2$	30.6	29.2
[^{12}C,^{13}C]C$_2$H$_2$	48.5	49.6
[^{13}C]C$_2$H$_2$	20.9	21.2

To account for the essential complete interchange of atoms, both C and H, a chain-reaction mechanism is suggested. The chain is initiated by the fragmentation of a relatively few acetylene molecules into C$_2$H, C$_2$, CH, and H species. These fragments then collide with C$_2$H$_2$ molecules, exchanging atoms and splitting off additional fragments. Allowing a residence time of 0.1 msec at an average plasma temperature of 3000°K we can estimate that each molecule will undergo approximately 2×10^4 collisions. If an efficiency of only 10% is assumed, each molecule will experience 2000 viable collisions in the first inch of plasma. As the reaction mixture cools downstream the number of collisions will decrease and the chain reactions will terminate as two CH fragments or a C_2H and H fragment collide to reform an acetylene molecule.

Acknowledgments

The technical contributions of Albert Bothe, Robert McDonough, and Richard Ruediger in planning and performing the described experiments and the stimulating discussion with David Stickler concerning the gas dynamics of the process are acknowledged and appreciated. The mass spectrographic analyses were performed at the Massachusettts Institute of Technology on the high-resolution mass spectrograph. The guidance and assistance of Klaus Bieman and Guy Arsenault in analyzing the data were most valuable.

Literature Cited

1. Littlewood, K., McGrath, I. A., *Intern. Conf. Coal Sci.*, 5th, Sheltennam, 1963, paper C9.
2. Rau, E., Seglin, L., *Fuel* (1964) **43**, 147.
3. Bond, R. L., Ladner, W. R., McConnel, G. I. T., ADVAN. CHEM. SER. (1966) **55**, 650.
4. Kawa, W., Graves, R. D., Hiteshue, R. W., *U.S. Bur. Mines, Rept. Investigation* **6829**, 1966.
5. Kawana, Y., Makino, M., Kimura, T., *Kogyo Kagaku Zasshi.* (1966) **69**, 1144.
6. Krukonis, V. J., Schoenberg, T., *Intern. Conf. Coal Sci.*, 7th, Prague, 1968, paper 5-132.
7. Wagman, D. D., Kilpatrick, J. E., Pitzer, K. S., Rossini, F. D., *J. Res. Nat. Std.* (1945) **35**, 467.
8. Howard, J., Wood, R., Kaltenbacher, D., *Chem. Eng. Prog.* (1961) **57** (11), 50.
9. JANAF Thermochemical Data Compiled and Calculated by the Dow Chemical Co., Thermal Laboratory, Midland, Mich.
10. Krukonis, V. J., Gannon, R. E., Schoenberg, T., *Can. Eng. Conf.*, 19th, Edmonton, Alberta, 1969.
11. Gannon, R. E., Krukonis, V. J., "Abstracts of Papers," 160th National Meeting, ACS, Sept. 1970, FUEL 44.
12. Plooster, M. N., Reed, T. B., *J. Chem. Phys.* (1959) **31**, 66.
13. Baddour, R. F., Iwasyk, J. M., *Ind. Eng. Chem., Process Des. Develop.* (1962) **1**, 169.
14. Baddour, R. F., Blanchet, J. L., *Ind. Eng. Chem., Process Des. Develop.* (1964) **3**, 258.

RECEIVED May 25, 1973. Work supported by the Office of Coal Research, U.S. Department of the Interior, under contract No. 14-01-000-493.

4

Arc Synthesis of Hydrocarbons

CHARLES SHEER and SAMUEL KORMAN

Chemical Engineering Research Laboratories, Columbia University,
632 W. 125th St., New York, N. Y. 10027

A convective arc featuring a novel cathode injection system has been studied on a laboratory scale producing hydrocarbons. Injection gases included H_2, $CO-H_2$ mixture, and steam projected against a carbon anode. Also injected was a powdered solid, $(CH_2)_x$, entrained in argon. The arc effluent was withdrawn via a hole in the anode through an intermediate hot zone to sampling equipment. A spectrum of operating parameters was studied whereby the hydrocarbon in the product could be varied from pure methane to pure acetylene. A catalytic surface effect on effluent composition within the intermediate hot zone was also observed involving wall temperature, contact surface material, and residence time. The results indicate that this technique may ultimately be applied to coal gasification.

This report summarizes some results of an investigation using a novel type of arc for hydrocarbon synthesis. The principal reacting system was carbon and hydrogen. Substitutions for hydrogen were also employed in a limited way, including a 50-50 vol % mixture of CO and H_2, steam, and a solid petroleum residue, essentially $(CH_2)_x$.

The type of arc employed was developed in our laboratory and is characterized by a number of unique features; an important one is the high rate of continuous through-put of feed material including fluids. The feed is raised to high temperatures and comprises the plasma environment of the discharge, particularly the arc-conduction column maintained between the electrodes. The composition of the plasma is derived from that of the feed; however, the atomic, molecular, or free radical plasma species differ significantly from the molecular composition of the feed.

Conventional arcs, consisting of a gaseous, electrical conducting column joining a positive anode and a negative cathode, are used as a source of heat, for example, as in electric smelting processes. Heat is transferred to feed materials by radiation and conduction from the hot column. This is the zone of primary energy dissipation in which the electrical energy is converted to radiant energy and sensible heat which flow out in all directions through the intervening layer of atmosphere. The maximum temperature which may thereby be maintained in the surrounding charge is limited to about 2500°C. In such arcs, little if any of the material treated enters the energy dissipating region within the arc-conduction column.

The first opportunity for treating materials continuously to temperatures higher than 2500°C arose with the discovery by Beck (*1*) in 1910 of the high intensity arc. With this type of discharge, the substance of the anode can be vaporized and passed through the conduction column where the temperature is raised to 10,000°K or more (*2*). When the feed material is incorporated into the anode, a major fraction can be exposed to the high energy density of the column (*3*). This is suitable however only for treating solids.

Problems are encountered when one or more of the reactants is a gas. Foremost of these is the instability induced in the arc column by appreciable forced convection. A considerable effort has been expended during the past two decades in stabilizing arcs subject to vigorous convection (*4*). A number of stabilization techniques have evolved including confinement of the column in a water-cooled channel (*5*), vortex stabilization (*6*), and magnetic stabilization (*7*). In several techniques (*8, 9*) the problem is avoided to some extent by mixing a gaseous reactant with the arc effluent; then the column proper is not subject to strong convection. None of the above, however, achieves a high degree of penetration of the gas into the primary energy dissipation zone within the column.

Recent work in this laboratory showed that large quantities of gases can be injected into the arc column in a practical manner. The gas is injected from the cathode end by means of a specially designed annular nozzle surrounding the cathode. This device is called the fluid convection cathode (FCC).

Basis of the Fluid Convection Cathode

The arc column converges to a small tip at the cathode surface (*see* Figure 1). This convergence, representing an inhomogeneous electric current flux, defines a zone of inhomogeneity in the accompanying magnetic field that produces a fluid mechanical thrust away from the cathode toward the anode, thus causing a pressure gradient away from the cathode tip. To stabilize this gradient, gas is aspirated into the arc column in the region of inhomogeneity and is propelled away from the cathode, creating the cathode plasma jet (*10*). This region is the only portion of the arc other than the anode crater through which appreciable quantities of gas may be injected without disturbing the stability of the discharge.

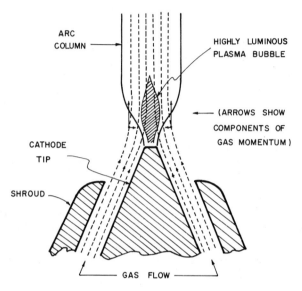

Figure 1. Sketch of FCC showing compressive effect of gas flow on base of arc column

The FCC was developed during a study of the influence of gas convection into the base of the arc column near the cathode of the arc (*11*). The conical tip of the cathode is surrounded by an annular nozzle which terminates upstream from the tip and which directs the gas in a converging high-speed layer into the column of the arc close to the point where it originates on the cathode. It was found that if this were done so that the gas impinged on the arc column in the contraction zone, the gas would preferentially enter the column. Further, the gas could be so injected at 10-20 times the natural aspiration rate. If an attempt is made to force the gas into the arc column elsewhere, the degree of penetration is far less and the injected gas tends to unstabilize and blow out the arc.

In confined arcs, where stability is achieved by enclosing the arc discharge within a water-cooled channel, it has been shown that over 70% of the injected gas never enters the column (*12*) and receives considerably less than the maximum possible activation energy from the arc.

Experimental

Figure 2 is a diagram of the arc apparatus, showing an FCC cathode through which the hydrogen gas is injected into the conduction column and a 1-inch diameter cylindrical carbon anode. The anode has a 1/4-inch hole along its longitudinal axis. The anode is connected at its back end by 3/16-inch id metal tubing to a 11/16-inch diameter, type 304 stainless steel tube surrounded by an electrically heated laboratory tube furnace. Leaving the furnace is a water-cooled heat exchanger following a tee connection valved to permit the gas stream to vent in either of two directions: (1) to a flowmeter and laboratory pump, or (2) to a manifold of a vacuum gauge and several valved 500-ml gas-sampling bottles. These are evacuated before use. The carbon anode can thus serve as a combination source of solid carbon and of carbon vapor issuing from the anodic arc terminus into the plasma column, as well as an arc crater gas-sampling probe.

Depending on the pumping flow rate or timed pressure rise in the sampling branch, it is possible to draw an arc flame effluent gas stream from the reaction zone through the tube furnace to vary the residence time at any temperature up to about 1000°C and thence through the heat exchanger and into the sample bottles in sequence.

In operation with diametrically opposed electrodes, the FCC arc column bears directly on the carbon anode which is completely covered by the arc crater at 150 amp or more. The pump valve is opened sufficiently to meter the plasma down through the anode hole, to purge and

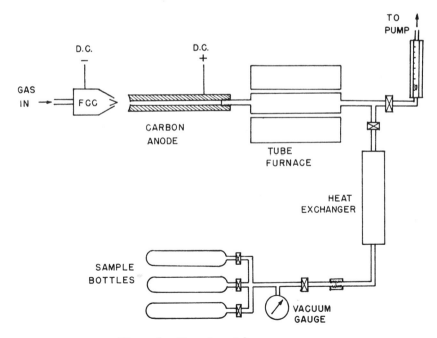

Figure 2. Experimental arrangement

equilibrate the effluent hot zone in the tube furnace, and then to meter samples into the gas sampling bottles in sequence at timed rates, measuring the rise period of the vacuum-gauge pressure.

The samples were analyzed by gas chromatography using helium carrier gas and an air-hydrogen flame in a Model 609 F&M Scientific Corp. flame ionization chromatograph with a Poropak Q column. A test gas mixture containing the aliphatic compounds CH_4, C_2H_2, C_2H_4, C_2H_6, C_3H_6, and C_4H_8 was used to calibrate the analytical procedure.

Results

The early objectives of the program were primarily exploratory, so the results reported here are essentially qualitative although within a given test series weight can be given to concentration ratios of components of a test sample mixture for comparison purposes.

Series I. Hydrogen Flow Rate Through the FCC. This was effected by operating the arc at standard conditions of 150 amp, maintaining the effluent hot zone at 800°C, sampling at about 30-60 sec per sample, and varying the hydrogen flow rate into the FCC. Comparison of hydrocarbon composition is shown in Table I in terms of the relative distribution of the volume concentration of the products found. The distribution was obtained by calculating the per cent contribution which each chromatographic amplitude recording made to the sum of all, in arbitrary scale divisions. There appears to be a significant dependence of effluent hydrocarbon composition upon the amount of hydrogen fed into the FCC.

Table I. Relative Distribution in Effluent *vs.* H_2 Flow Through FCC

H_2, moles/min	CH_4	C_2H_2	C_3H_6
3.4	0	100	0
6	4.4	66.2	29.4
8.5	45.6	45.6	8.8
14.1	85	15	0

Series II. Time Factor. Standard conditions of 150 amp and 8.5 moles hydrogen/min were used with varying sampling rates through the 800°C effluent hot zone. Results are given in Table II. No other hydrocarbons were observed. These data suggest that methane and acetylene are produced and disappear at different rates.

Series III. Hot Zone Temperature. Standard conditions included 150 amp, 8.5 moles hydrogen/min through the FCC, and sampling rate through the effluent hot zone at 2½ min with variation of the hot zone temperature. Results are shown in Table III. It is evident that the hot zone temperature has a significant effect on the hydrocarbon composition of the effluent.

Table II. Relative Distribution vs. Sampling Flow Time Through Hot Zone

Time	CH_4	C_2H_2
10 sec	44	56
12	75.5	24.5
20	91.5	8.5
30	95.5	4.5
1¼ min	93	7
2½	100	0
4	9	0
12	0	0

Table III. Relative Distribution vs. Hot Zone Temperature

Temp., °C	CH_4	C_2H_2	C_3H_6
800	100	0	0
500	35	65	0
400	40	60	0
200	2	87.5	10.5
25	0.5	99.5	0

Series IV. Hot Zone Surface Area. We noted the result of increased time of flow in the sampling rate shown above in Series II and assumed that 8.5 moles hydrogen/min through the FCC creates a steady state for carbon and hydrogen in the plasma at the arc crater (the sampling source), then the time of exposure to the hot zone wall of type 304 stainless steel was observed. This was accomplished at 150 amp, 8.5 moles hydrogen/min, and 800°C hot zone temperature, in two diameters of hot zone tubes, 11/16 and 9/16 inch—a cross-sectional area ratio of 4:1. To equate the sample residence times, the sample flow periods were adjusted to this ratio. Results are shown in Table IV. No acetylene was found in the 9/16-inch diameter samples while the small amounts in the 11/16-inch samples were consistent with the distribution for 20 and 30 scc checked with similar times observed in the earlier Series II above.

Table IV. Hydrocarbon Ratio vs. Hot Zone Surface

Diam.	Sampling Rate, sec	CH_4 Ratio
9/16:1-1/16	80:20	1:11
	120:30	1:8

Series V. Further Effect of Hot Zone Area. The result of Series IV was followed by further observations comparing hydrocarbon yields and ratios in two cases and at two temperatures, as follows:

A. 11/16-inch diameter at 800° and 500°C

B. 11/16-inch diameter, into which tube a section of stainless steel wool was added, also at 800° and 500°C

Table V. Hydrocarbon Distribution vs. Surface Area

Sampling Time, sec	800°C				500°C			
	Without St. St. Wool		With St. St. Wool		Without St. St. Wool		With St. St. Wool	
	CH_4	C_2H_2	CH_4	C_2H_2	CH_4	C_2H_2	CH_4	C_2H_2
30	94	6	100	0	18	82	84	16
150	100	0	100	0	35	65	100	0

These conditions were compared because any effect resulting from exposure to a large surface of stainless steel would not require comparably rapid sampling flow. In other words, the effect of increased surface area alone could be observed. The hydrocarbon distribution at each temperature with and without added stainless steel wool is shown in Table V. Table V indicates that the preponderance of methane and absence of acetylene is not affected at 800°C. To interpret the apparent shift, however, at 500°C it is necessary to compare the relative concentrations of all samples. This is shown in terms of their ratios in Table VI. It will be noted that the effect of the stainless steel hot zone is relatively constant and appreciable for the times and temperatures of exposure. At 800°C, no acetylene is present, as expected for temperature as shown in Table III above while at 500°C the time-related suppression of acetylene previously observed in Table II above is also more strongly enhanced by the increased stainless steel surface area. We interpret this to mean that acetylene disappears much more rapidly than methane under these conditions and that the disappearance is related to the surface area of the stainless steel hot zone.

Table VI. Ratio of Hydrocarbon Concentrations in Samples

Sampling Time, sec	800°C Ratio, +/−[a]		500°C Ratio, +/−[a]	
	CH_4	C_2H_2	CH_4	C_2H_2
30	1/1.3	0/0	1/1.7	1/27
150	1/4	0/0	1/4.8	1/200

[a] + = with stainless steel wool; − = without stainless steel wool.

Series VI. Effect of Hot Zone Surface Composition. We noted that the time-related suppression of methane and acetylene suggested a hot zone surface effect when stainless steel was used, so this material was replaced by several others, using 11/16-inch diameter tubes. Arc crater gas samples taken under otherwise identical conditions (viz., 150 amp, 8.5 moles hydrogen/min, hot zone temperature 800°C, parallel sampling flow rates) produced hydrocarbon compositions as shown in Figures 3-6.

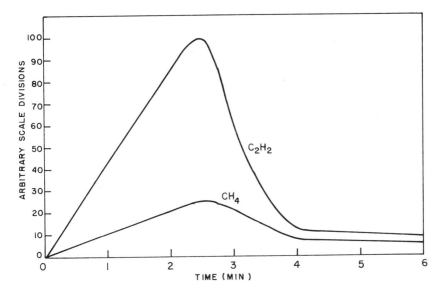

Figure 3. Silica

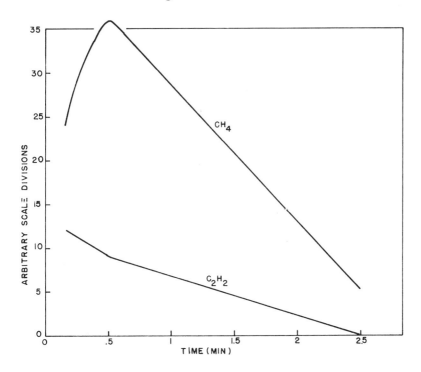

Figure 4. Iron

For fused silica (Figure 3), the time of exposure in the 800°C hot zone has roughly parallel effects on the presence and disappearance of both methane and acetylene. This is in contrast with iron (Figure 4) and type 304 stainless steel (Figure 5) where acetylene disappears rapidly while methane persists. Increasing the nickel content by using Incoloy 800 (32% Ni, 46% Fe) and nickel-200 (99.5% Ni) appears to produce further suppression of both acetylene and methane—an effect which does not appear to be especially time-sensitive.

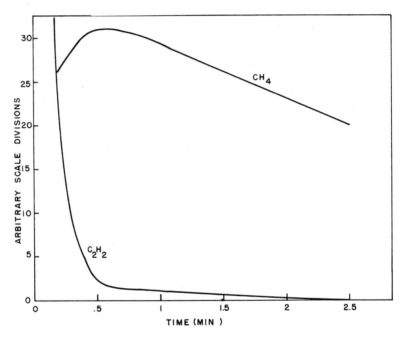

Figure 5. Stainless steel type 304 (8% Ni, 74% Fe)

Some Results with Other FCC Gas Feeds. Preliminary tests were carried out in which substitution was made for hydrogen as the FCC-injected gas. The first substitute was a H_2-CO, 50-50 vol % mixture. The usual standard test conditions were employed, except that the gas volume flow rate was set to a value which included a relatively small amount of hydrogen (0.6 mole/min). The results (Table VII) show a comparison of the (interpolated) analog distribution resulting from the same amount of pure hydrogen alone with the distribution using the mixture with CO. The absolute amount of acetylene in the CO-H_2 mixture tests also increased with time whereas the acetylene in the low-rate pure hydrogen analog diminished rapidly and no hydrocarbon was found after 10 sec.

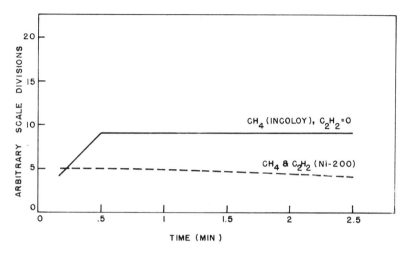

Figure 6. *Incoloy 800 (32% Ni, 46% Fe); Ni-200 (99.5% Ni)*

Table VII. **Hydrocarbon Distribution: (CO + H_2) vs. H_2**

Sample Time	$(CO+H_2)$ $CH_4:C_2H_2:C_3H_6$	H_2
10 sec	30:60:10	100% C_2H_2
30	40:60	0
2½ min	16:81:3	0

The next FCC gas used was steam, with corresponding results shown in Table VIII. In a comparison of absolute concentrations by measuring chromatographic deflection amplitudes, the total product yield of hydrocarbon increased with increasing flow rate of steam. As with the water-gas tests, short time intervals with steam produced appreciable methane under conditions which would have yielded only a small amount of acetylene or no hydrocarbons if the same amount of contained hydrogen were fed as pure hydrogen in equivalent rates to the FCC arc.

Table VIII. **Hydrocarbon Distribution Ratio for FCC Steam + Carbon Anode**

Steam, grams/min	Contained H_2, gram/min	Sample Time, sec	Distribution Ratio, $CH_4:C_2H_2$
2	0.22	30	5:4
2	0.22	90	3:5
4	0.44	10	8:1
4	0.44	30	CH_4 only
6.3	0.7	10	4:1

Solid Carbonaceous Feed. Finally, preliminary tests were carried out in which a powdered solid was entrained in argon and injected into the FCC arc. The solid was a petroleum refinery bottom having a softening point of 327°F and an approximate composition of $(CH_2)_x$. The argon was chosen as a neutral carrier to avoid ambiguity concerning the source of hydrogen in the product. For simplicity, no effluent hot zone was used, so the sample gas was considered to be quenched to room temperature. As expected, in every instance the hydrocarbon product was preponderantly acetylene although there were traces of methane also.

Discussion

Based on the results of the tests described above, two distinct sets of processes are operative. One involves the phenomena within the arc and the other relates to the conditions to which arc-generated gases are exposed in the effluent stream.

This investigation is exploratory, so it affords no evidence which discloses the undoubtedly complex mechanisms underlying the observed effects. However, it seems reasonable to assume that a time-dependent catalytic effect exists which is related to the composition and temperature of the hot zone surface to which the arc sample effluent was exposed. The rapid transition in hydrocarbon composition from acetylene to methane in the presence of iron or stainless steel is one indication. Thermal effects and increased residence time in the presence of hydrogen lead to progressive diminution and disappearance of hydrocarbons, suggesting another, slower process which may be pyrolytic or possibly inhibitory.

It is submitted, also, that effects which occur within the hot zone are not necessarily independent of the arc reactions. The composition of the effluent which leaves the arc crater and enters the hot zone has an important effect on the ultimate product composition. For example, the contrast, under otherwise identical test conditions, which is evident as substitution was made for pure hydrogen as the FCC gas feed produced a noteworthy change. With a mixture of CO and H_2, or H combined with O as steam, hydrocarbons were produced with appreciable or major fractions of methane while pure hydrogen yielded only small amounts of acetylene or no hydrocarbons at all. One may infer that the presence of CO or O within the plasma possibly alters the course of the reaction and hence the effluent composition.

From our basic studies of FCC gas injection (*13*) into the arc-conduction column, together with concurrent temperature measurements, it is fairly certain that about 80% of the injected gas penetrates the column and reaches temperatures greater than 10,000°K. Our current investigations of this arc, using hydrogen as FCC gas feed and a carbon anode, indicate clearly that the hydrogen in the arc-conduction column approaching the anode is monatomic. We are presently studying the plasma composition in front of the anode crater, where this monatomic hydrogen is mixing with or impinging on carbon at its sublimation temperature. The objective of identifying the plasma species in this zone, with hydrogen and ultimately with H_2 and CO or with steam, offers the possibility of determining how the ultimate quenched effluent composition, as well as the function of the secondary species such as CO or O in altering the process, may be predicted. In addition the exposure of a carbonaceous solid into this plasma, producing acetylene in accordance with expectation, indicates in a preliminary way the possibility of employing these arcs effectively for gasification involving coal or other carbonaceous feeds.

Literature Cited

1. Beck, H., *Elektrotech. Zeit.* (1921) **42**, 993.
2. Finkelnburg, W., "Der Hochstromkohlebogen," Springer-Verlag, Heidelberg, 1948.
3. Sheer, C., Korman, S., in "Arcs in Inert Atmospheres and Vacuum," W. E. Kuhn, Ed., p. 175, John Wiley and Sons, New York, 1956.
4. Sheer, C., in "Vistas in Science," D. L. Arm, Ed., p. 135, University of New Mexico Press, Albuquerque, 1968.
5. Reed, T. B., in "Advances in High Temperature Chemistry," L. Eyring, Ed., sec. IV, Academic, New York, 1967.
6. Gladisch, H., *Hydr. Proc. Pet. Ref.* (1962) **41**, 159.
7. Schultz, R. A., unpublished report of Dupont Corp., Feb. 22, 1968.
8. Gladisch, H., *Chem. Ing. Technik.* (1969) **41**, 204.
9. Gannon, R. E., Krukonis, V., *Avco Corp. Rept. No.* **34**, O.C.R. Contract No. 14-01-0001-493, April 1972.
10. Maecker, H., *Zeit. Physik.* (1955) **141**, 198.
11. Sheer, C., Korman, S., Stojanoff, C. G., Tschangs, P. S., *Rept. No.* **AFOSR 70-0195TR**, Aeromechanics Directorate, Air Force Office of Scientific Research, Arlington, Mar. 1969, p. 33.
12. Emmons, H. W., in "Modern Developments in Heat Transfer," W. Ibele, Ed., p. 465, Academic, New York, 1963.
13. Sheer, C., Korman, S., Kang, S. F., *Final Rept., Contract No.* **F44620-69-C-0104**, Aeromechanics Directorate, Air Force Office of Scientific Research, Arlington, in press.

RECEIVED May 25, 1973. Work supported by Consolidated Natural Gas Service Co. Basic investigation of fluid convection cathode sponsored by Aeromechanics Directorate, Air Force Office of Scientific Research, Contract No. F 44620-69-C-0104.

5

The Reaction of Atomic Hydrogen with Carbon

ALAN SNELSON

IIT Research Institute, Chicago, Ill. 60616

A thermally produced beam of atomic hydrogen was allowed to react on a carbon target at temperatures between 30° and 950°C. The reaction products were isolated on a liquid helium cold finger and then analyzed by gas chromatography. Over the temperature range examined the major reaction products were: 91% CH_4, 8.4% C_2H_6, 0.6% C_3H_8. C_2H_4, C_3H_6, and C_4 hydrocarbons were minor constituents, if formed at all. Hydrocarbon formation increased with temperature; there was no maximum in the yield occurring at about 770°K as reported in previous studies. The temperature dependence of the methane yield showed three distinct phases, and activation energies were obtained. At 30° and 950°C, about 1 and 3%, respectively, of the available H atoms reacted with carbon to form hydrocarbons.

Aramenko discovered in 1946 ([1]) that H atoms react with carbon. To date, the results of 10 other studies have been reported ([2, 3, 4, 5, 6, 7, 8, 9, 10, 11]). In all but one investigation, H atoms were produced by electric discharge techniques; the one exception ([7]) specified thermal methods. The reaction products were analyzed chemically in all but one study ([5]). Agreement among the different investigations as to the hydrocarbons formed in the H atom–carbon reaction is not good; some or all of the following have been reported: CH_4, C_2H_2, C_2H_4, C_2H_6, C_3H_8, and various butanes and butenes. Several authors believe that methane is the primary reaction product, with higher hydrocarbons resulting from hydrogen-abstraction reactions and free radical-recombination processes. There is some indication that the formation of acetylene and ethylene may be associated with ionic species formed in the electric discharge used to produce the atomic hydrogen. In some of the experimental studies the products reported as being formed in the H atom–carbon reaction could also have been formed by H atom attack on organic materials, *e.g.*, vacuum grease and O-rings, which were part of the system. Most investigations of the H atom–carbon reaction were made at ambient temperatures, but in two cases (8, 9) where an extensive temperature range was investigated, a maximum in the hydrocarbon yield was reported at 770° ± 50°K. In two kinetic studies (6, 9) on the reaction, data were obtained for the rate of carbon removal as a function of temperature and H atom concentration, but no effort was made to correlate these data with the hydrocarbon production.

In this investigation the H atom–carbon reaction has been reexamined in an effort to determine 1) the nature of the hydrocarbon products, 2) the yields of hydrocarbon products as a function of temperature and, 3) the efficiency of conversion of atomic hydrogen to hydrocarbons.

Experimental

In designing the experimental arrangement for studying the H atom–carbon reaction, an effort was made to avoid some of the features which may have vitiated the results obtained in previous studies, *e.g.*, possible pyrolysis of the reaction products, reaction between H atoms and O-rings or vacuum grease, and reaction between hydrogen ions and carbon. To attain these goals, a low pressure, atomic hydrogen beam–carbon reactor was constructed using a liquid helium cold finger to remove reaction products. This is shown schematically in Figure 1. All materials used in the construction of the reactor were either metal (copper, brass, Kovar, and stainless steel) or glass. Demountable joints were soft-soldered using an inorganic flux, and all surfaces were acid-cleaned prior to assembly. A mechanical vacuum pump and an oil-diffusion pump were used to evacuate the system, and pressures in the 10^{-6} mm range were routinely achieved.

Hydrogen atoms were formed by thermal low pressure (10^{-7}–10^{-5} atm) dissociation of molecular hydrogen in a tungsten effusion tube. The effusion tube, 0.067 inch od and 0.030 inch id, was heated for over 2 inches to 2600° ± 50°K by electrical induction. The temperature was measured with a Leeds Northrop optical pyrometer; emissivity corrections were made. To increase the probability of attaining equilibrium within the tungsten effusion tube for the chemical reaction $\underset{\text{temp}}{\overset{\text{high}}{\rightarrow}} 2H$, three tungsten wires about 1 inch long and of 0.010 inch diameter were inserted into the bore of the tube to help increase residence times. Matheson research grade hydrogen was used in the study and was stored in a glass vacuum line prior to use. The flow rate of hydrogen to the effusion tube was controlled by varying the gas pressure across a fixed leak. Hydrogen flow rates to the effusion tube varied from 0.5 to 9×10^{-5} mole/hr. In most experiments the rate was 2×10^{-5} mole/hr. The amounts of gas being fed to the reactor were determined from pressure-volume-temperature data using standard vacuum line techniques. These hydrogen flow rates correspond to pressures of about 5×10^{-7} to 2×10^{-5} atm at the lowest and highest flow rates, respectively. At the most commonly used hydrogen feed rate (2×10^{-5} mole/hr, the pressure in the effusion tube was approximately 4×10^{-6} atm. According to data given in the JANAF Tables (12), the resulting effusate at this pressure is essentially pure atomic hydrogen. At the highest experimentally used effusion tube pressure (2×10^{-5} atm) the effusate was $> 98\%$ atomic hydrogen.

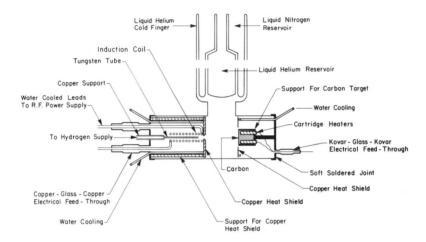

Figure 1. Schematic arrangement of atomic hydrogen–carbon reaction cell

The carbon used in the H atom–carbon reaction was obtained from the Ultra Carbon Corp., Mich. It had a certified purity of 99.9995% and a density of 1.72 gram/cm^3 and was in the form of a solid cylinder, 1 inch long $\times$ 1 inch diameter. In the reactor, the carbon was mounted in a steel holder on the same axis as the tungsten effusion tube, 2.63 inches from it, presenting the H atoms with a flat target surface of 1 inch diameter. The carbon target was heated by four Waltow cartridge heaters of 100 W each, and the temperature was controlled by varying the applied voltage across the heaters. A chromel-alumel thermocouple, inserted into the carbon target with the temperature sensing junction about 1/16 inch from the reaction face, was used to measure the target temperature. The thermocouple output was measured against that of an ice junction using a Leeds Northrop potentiometer. Target temperatures were constant to $\pm 2°C$ in experiments which usually lasted 1 hr. A maximum target temperature of about 1000°C was possible, and prior to use in the H atom–carbon reaction the carbon was outgassed at 950°C for three days under vacuum.

The glass liquid helium cold finger used to freeze out reaction products was about 2 inches in diameter and had a 1 liter capacity. To minimize heat leakage it was surrounded by a liquid nitrogen heat shield. All surfaces in the Dewar system were silvered. After an experiment the cryogenic fluids were removed from the cold finger, and the system was warmed to room temperature. The reaction products were then removed by a Toepler pump, and their total volume was measured. During this process the carbon target was maintained at 200° ± 20°C to help minimize adsorption of gas. Provision was made in the collection section of the vacuum line to raise the pressure of the collected sample to slightly above ambient. This was done to help improve the reliability of the gas sampling for chromatographic analysis.

Gas samples from the H atom–carbon reaction were analyzed on a Varian-Aerograph gas chromatograph model 1800. A 6-ft long, ¼-inch diameter stainless steel column packed with 216 grams of Poropak Q stationary phase was used for separating the various hydrocarbons. The column was used isothermally (60°C) for the analysis of CH_4, C_2H_4, and C_2H_6. For higher hydrocarbons the column temperature was programmed at 10°/min for 6 min and then held steady at 120°C. Under these conditions 10^{-11} mole of a simple hydrocarbon could be detected. Before every analysis the calibration of the chromatograph was checked against injections of known volumes of methane and ethylene. In a typical experiment, 0.15 ml of the collected gas from the atomic hydrogen–carbon reactor was injected into the chromatograph, resulting in the following hydrocarbon yields: 4-18 × 10^{-8} mole CH_4, 1-2 × 10^{-9} mole C_2H_4, 3-7 × 10^{-9} mole C_2H_6, 0.5-2 × 10^{-10} mole C_3H_6, and 2-5 × 10^{-10} mole C_3H_8. The overall accuracy of the measured hydrocarbon yield was about ±7%.

Results and Discussion

Operating Characteristics of the Reactor without the Carbon Target. The reactor, without the carbon target in place, was exposed to H atom attack using a hydrogen flow rate to the reactor of about 2 × 10^{-5} mole/hr. The resulting product gases were analyzed chromatographically. The following species were found: CH_4, C_2H_4, C_2H_6, C_3H_6, C_3H_8, and traces of butanes and butenes. These findings were surprising because great efforts were made to remove organic materials from all reactor surfaces before assembly. The reactor was disassembled, and all surfaces were inspected and recleaned. On re-assembly and subjection to H atom attack, hydrocarbons were again produced. The effect of exposing the reactor to H atom attack over prolonged periods of time was therefore studied. Samples of the reaction products were analyzed periodically.

Data obtained from these experiments are given in Table I. To permit comparison between different experiments, reaction product yields are all quoted in terms of moles of hydrocarbon formed per mole of molecular hydrogen fed to the reactor at the stated molecular hydrogen feed rate. Quantitative yields for the C_4 hydrocarbons are not given in Table I because their amounts were too small for meaningful analysis.

Table I. Experiments to Determine Background Hydrocarbon Yield

Expt.	H_2 Feed Rate, mole $\times$ 10^{-5}/hr	Mole $\times$ 10^{-3} Hydrocarbon Formed CH_4	C_2H_4
52	1.9430	9.4765	1.2666
53	1.9350	3.0493	0.3192
54	1.8917	1.6025	0.0890
55	2.1736	2.2031	0.1710
56	1.9635	1.7013	0.1188
57	1.9293	1.8502	0.1024
58	2.0461	1.2485	0.1344
59	2.0885	1.2118	0.1719
60	1.9907	1.6453	0.0809
61[a]	0.5061	4.0439	0.6047
62	2.1318	2.0290	0.0813
63[a]	0.7561	2.5788	0.2263
64[a]	9.7126	0.3339	0.0127
65[a]	4.0310	0.5501	0.0401
66[a]	6.5012	0.7081	0.0220

[a] These data points were used in establishing the curve shown in Figure 3.

In Figure 2 the methane yield as a function of time is presented from the data given in Table I, with a hydrogen feed rate to the reactor of 2×10^{-5} mole/hr. It is at once apparent that the yield of methane decreases quite markedly with the number of hours of H atom attack (conditioning), and after 60-80 hrs it appears to reach a constant minimum. The production of C_2H_4, C_2H_6, C_3H_6, and C_3H_8 showed the same type of behavior as methane with respect to yields as a function of H atom conditioning time. After 60-80 hr a stable minimum was attained for all species. The reactor was disassembled and inspected, surfaces were cleaned, and the reactor was re-assembled. The hydrocarbon yield as a function of H atom-conditioning time was re-investigated. The same type of behavior resulted as in the prior experiments—a moderate initial hydrocarbon yield, decreasing with time after 60-80 hrs to the same value obtained previously. Conditioning was continued for a total of 150 hrs without any further change in the hydrocarbon yield. Table II lists the stable hydrocarbon yields obtained after prolonged H atom conditioning of the reactor.

From these data it is necessary to conclude that despite all efforts to maintain reactor cleanliness, the system was contaminated with carbon or organic material. The initial relatively large production of hydrocarbons obtained directly after assembling the reactor could have resulted from hydrogen atom attack on freshly adsorbed carbon species, CO, CO_2, and possibly some hydrocarbons, on the inside of the reactor after exposure to the laboratory atmosphere. The apparently smaller constant yield of hydrocarbons obtained after prolonged hydrogen atom attack suggests the presence of a fairly large though not particularly accessible supply

H Atom Reaction with Residual Carbon or Carbon Species in Reactor

C_2H_6	C_3H_6	C_3H_8
0.7516	0.0983	0.1629
0.1939	0.0147	0.0276
0.1266	0.0057	0.0240
0.1711	0.0114	0.0303
0.1406	0.0076	0.0267
0.1196	0.0091	0.0233
0.1217	0.0055	0.0192
0.1142	0.0056	0.0206
0.1457	0.0092	0.0258
0.4110	0.0170	0.0810
0.1960	0.0057	0.0337
0.3853	0.0090	0.0677
0.0315	0.0005	0.0052
0.0310	0.0045	0.0052
0.0750	0.0005	0.0049

of organic material. There are two possible sources: carbon in the steel used in fabricating some parts of the reactor or organic material trapped during the formation of the silver reflective coating on the liquid helium Dewar. (The latter coatings are prepared by the reduction of ammonical silver solutions with sugar.) In view of the large silvered surface area of the liquid helium Dewar, about 5000 cm², a small amount of trapped organic material in the silver coating could well be the major source of organic contamination in the reactor.

After establishing that a small constant yield of hydrocarbons could be obtained from the reactor on hydrogen atom attack, a few experiments were tried in which the hydrogen feed rate to the reactor was varied from 0.25 to 9.75 × 10⁻⁵ mole/hr, and the reaction products were analyzed. These data are recorded in Table I. In Figure 3, the methane yield as a function of the hydrogen feed rate is presented, based on data given in Table I. Similar curves also resulted for the other hydrocarbons, and they are not presented individually. From these data, hydrocarbon yields vary by a factor of about eight with respect to hydrogen feed rate, increasing at the lower feed rates and decreasing at the higher feed rates. As noted in the experimental section, over this range of feed rates the hydrogen species leaving the effusing tube is essentially pure atomic hydrogen with at most a 1-2% variation occurring between the lowest and highest flows. Such small changes in the H atom concentration cannot explain the observed variation in hydrocarbon yield as a function of feed rate. At lower hydrogen feed rates, H atom recombination reactions will occur at a lower rate than at higher hydrogen feed

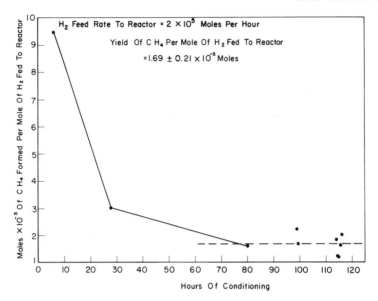

Figure 2. Yield of CH_4 in reactor without carbon target as a function of H atom conditioning time

rates. It is possible that the decreased H atom recombination rates at the lower feed rates, and hence longer H atom life time, increases the probability for H atom surface reactions producing hydrocarbons. A more detailed study of the effect was not undertaken.

A few experiments were tried to determine if significant cracking of hydrocarbons occurred in the reactor on the hot effusion tube. To this end small quantities of methane were introduced into the reactor and allowed to impinge on the carbon target before being frozen out on the liquid helium cold finger. The products were subsequently analyzed. Within the sensitivity of the chromatographic detection, no noticeable cracking of the methane occurred.

Table II. Hydrocarbon Yield after Conditioning the Reactor for a Prolonged Period of Time (>80 hr) at a Hydrogen Feed Rate of 2×10^{-5} mole/hr

Hydrocarbon	mole Hydrocarbon/ mole H_2 fed to Reactor[a]
CH_4	$(1.69 \pm 0.21) \times 10^{-3}$
C_2H_4	$(1.18 \pm 0.27) \times 10^{-4}$
C_2H_6	$(1.42 \pm 0.20) \times 10^{-4}$
C_3H_6	$(7.5 \pm 1.6) \times 10^{-6}$
C_3H_8	$(2.55 \pm 0.34) \times 10^{-5}$

[a] Reported error limits are the standard deviations of the observed experimental values.

The above experiments served to characterize the operation of the H atom reactor. The apparent inability to remove all traces of organic material from reactor surfaces exposed to H atoms was not expected. It is interesting to note that in all previous experimental studies on the H atom–carbon reaction in which hydrocarbon reaction products were analyzed, no reports were made of tests to determine possible hydrocarbon yields in the absence of the carbon target. Sufficient experimental details were given in some of these studies to indicate that H atom attack on O-rings and vacuum greases in the reactor system probably occurred.

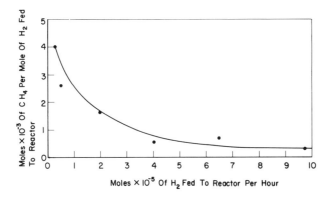

Figure 3. Yield of CH_4 as a function of H_2 feed rate

Hydrogen Atom–Carbon Reaction. The carbon target was placed in the reactor which was then conditioned to H atom attack for about 200 hrs. Experiments were then conducted to determine the hydrocarbon yield from the H atom–carbon reaction as a function of temperature over the range 30°–950°C. Hydrogen flow rates of 2×10^{-5} mole/hr were used in all experiments. The following hydrocarbons were detected: CH_4, C_2H_4, C_2H_6, and C_3H_6, with possibly minute traces of some C_4 species. The amounts of the latter were similar to those detected in the background yield without the carbon target present, and reliable quantitative measurements were not possible. Qualitatively, C_4 hydrocarbon yields were estimated at $< 10^{-3}$ of the methane. For the C_1, C_2, and C_3 hydrocarbons, yields for the H atom reaction with the carbon target were calculated by subtracting the background yield from the total in the sample. This method of calculating the hydrocarbon yield assumes that the background level of hydrocarbons is not affected by the presence of the competing process at the carbon target. This subject is discussed later.

The data obtained from 25 experiments are shown in Table III. The hydrocarbon yields are presented in terms of moles of hydrocarbon

Table III. Data for Hydrocarbon Yields in the Range 301–1222°K

Expt.	Target Temp., °K	CH_4 Yield[a] $\times 10^{-3}$	C_2H_4 Yield $\times 10^{-3}$
1	1137	5.950	0.0053
2	983	4.706	0.0081
3	804	4.354	0.0046
4	626	4.265	0.0483
5	428	3.876	−0.0021
6	1108	6.460	0.0374
7	508	4.000	−0.0068
8	1206	7.109	0.0501
9	1075	5.265	−0.0086
10	333	2.838	−0.0540
11	1101	5.519	0.0262
12	898	4.217	−0.0375
13	301	2.723	0.0183
14	1144	6.188	0.0319
15	370	3.293	−0.0604
16	352	3.350	−0.0718
17	305	2.580	−0.0040
18	945	4.470	−0.0565
19	482	4.171	−0.0129
20	770	4.163	−0.0364
21	675	4.130	0.0404
22	306	2.399	−0.0318
23	386	3.758	−0.0748
24	1222	6.814	0.0207
25	616	3.932	−0.0648

[a] Yields are presented in terms of moles of hydrocarbon formed, per mole of H_2 fed to the reactor.

formed per mole of molecular hydrogen fed into the reactor at the stated temperature. In Figures 4, 5, and 6 the data are shown graphically for methane, ethane, and propane, respectively. The scatter of the individual points on the methane curve shown in Figure 4 may be considered acceptable in terms of the expected experimental errors, but the scatter of the points on the ethane and propane curves is considerably larger. The reason for the increased scatter is not certain. It was observed that yields of ethane and propane obtained in a low temperature experiment performed directly after a high temperature experiment appeared to be significantly higher than the yield obtained when performing two low temperature experiments consecutively. This behavior suggests some type of hysteresis effect is occurring which results in these hydrocarbons being more slowly released during sample collection than the methane. It is probable that these two hydrocarbons are more strongly adsorbed by the surfaces in the reactor than the methane, and this might account for the rather poor precision of the data. It was not possible to examine this problem in more detail.

with a Hydrogen Feed Rate of 2 × 10⁻⁵ mole/hr

C_2H_6 Yield × 10^{-3}	C_3H_6 Yield	C_3H_6 Yield
0.595	0.0036	0.0482
0.331	0.0108	0.0252
0.294	−0.0025	0.0275
0.298	−0.0030	0.0276
0.221	−0.0036	0.0206
0.648	−0.0003	0.0400
0.275	−0.0029	0.0225
0.725	0.0012	0.0548
0.400	−0.0029	0.0246
0.264	0.0037	0.0148
0.580	−0.0037	0.0311
0.259	−0.0031	0.0173
—	—	—
0.644	—	0.0296
0.215	0	0.0072
0.199	0	0.0071
0.201	−0.0019	0.0255
0.218	−0.0032	0.0093
0.216	−0.0045	0.0170
0.208	−0.0016	0.0093
0.241	−0.0022	0.0158
0.240	−0.0042	0.0239
0.264	−0.0039	0.0122
0.536	−0.0023	0.0466
0.274	−0.0051	0.0100

The data presented in Table III for yields of ethylene and propene in the H atom–carbon reaction include many negative values. These arise from the mode of calculations of the yield which was explained earlier. The negative values indicate that smaller quantities of ethylene and propylene are being formed in the presence of the carbon target than in its absence, suggesting that the interaction of the H atoms with the carbon target reduces the background yield of these hydrocarbons. From the data presented in Table III for C_2H_4 and C_3H_6, it appears that their respective yields are not noticeably temperature dependent. Assuming this to be true, the average yield of C_2H_4 is −(0.009 ± 0.039) × 10⁻³ and C_3H_6 is −(0.0014 ± 0.0035) × 10⁻³ mole/mole H_2 fed to the reactor over the temperature range investigated. The error limits associated with these values are quite large, so it cannot be unequivocally stated that no ethylene or propene is formed during the H atom–carbon reaction. However the data do suggest values close to, if not, zero for yields of these two unsaturated hydrocarbons. To add further credence to this conclusion, at the end of the series of experiments

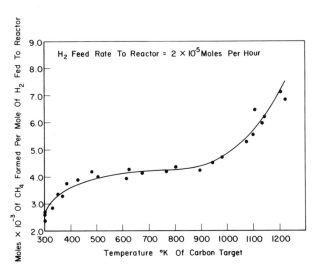

Figure 4. Production of methane as a function of temperature

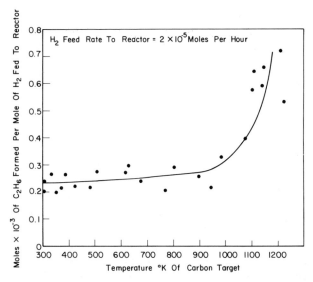

Figure 5. Production of ethane as a function of temperature

the carbon target was removed from the reactor and the background yield of hydrocarbons was again determined. After 80 hrs of conditioning the yields of hydrocarbons were all found to agree, within experimental error, with the values obtained previously.

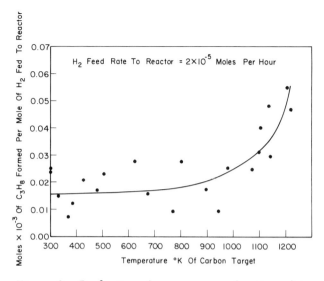

Figure 6. Production of propane as a function of temperature

In Figure 7 an Arrhenius plot of the H atom–carbon reaction data is presented based on the methane yields given in Table III. Three distinct reaction regions are indicated. A least-squares fit on the data resulted in activation energies for the production of methane of 4.5 ± 1.2, 0.15 ± 0.05, and 0.94 ± 0.20 kcal/mole in the high, medium, and low temperature regions, respectively. The precision limits are the standard deviations

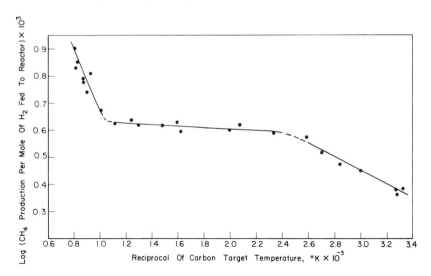

Figure 7. Arrhenius plot for methane production

calculated from the experimental data. No attempt was made to fit the data obtained from the ethane and propane yields to an Arrhenius-type curve because of their poor precision.

Hydrocarbon Product Distribution. The present investigation has established that the reaction of hydrogen atoms, at an initial temperature of about 2600°K, with a carbon surface at approximately 30°C produces the following saturated hydrocarbons: 92 mole % CH_4, 8.6 mole% C_2H_6, and 0.4 mole % C_3H_8. These ratios do not appear to vary significantly over the temperature range studied. There is a possibility that very minor amounts of C_2H_4, C_3H_6, and C_4 species may also be formed. In earlier studies on the atomic hydrogen–carbon reaction, Wood and Wise (9) reported hydrocarbon yields of 91% CH_4 with 9% C_2-C_8. Harris and Tickner (2) have reported 91% CH_4 and 9% C_2-C_5. In both these studies low pressure electric discharges were used to produce hydrogen atoms. In the former study, sufficient experimental details were given to indicate that ionic species from the electric discharge probably did not take part in the reaction. The possibility that some H atom attack on O-rings in the system occurred cannot be ruled out. The reported presence, though presumably small, of hydrocarbons in the C_4-C_8 range suggests, based on the results obtained in this study, that H atom attack on some material other than the carbon target occurred. Lack of experimental details does not allow an assessment of the validity of the results obtained in the Harris and Tickner study. In the study by Gill *et al.* (7), low pressure thermally produced hydrogen atoms reacting with various carbons were reported to produce methane yields from 89.4 to 49.8%, with higher hydrocarbons, C_2-C_4, making up the balance. In this study hydrogen atom attack on organic material within the reaction vessel undoubtedly occurred, and these results must therefore be discounted. It is not possible to compare the values obtained for the relative yields of the individual higher hydrocarbons determined in this study with those of earlier investigations because either no specific data were given, or in those cases where they were (9) the experimental procedures were apparently unreliable.

It has been suggested that CH_4 is the primary product in the H atom–carbon reaction and that higher hydrocarbons are the result of secondary processes. The latter could be surface or gas-phase interactions between H atoms and CH_4 or between radical species, CH_3, CH_2, or CH. In the present investigation, significant quantities of both C_2H_6 and C_3H_8 were formed in addition to CH_4, and although the experimental conditions probably eliminated secondary gas phase reactions, secondary surface processes could certainly have occurred. The possible primary or secondary nature of C_2H_6 and C_3H_8 in the H atom–carbon reaction cannot be determined from data obtained in this study. Had the precision of the yield data for both C_2H_6 and C_3H_8 been comparable with that for CH_4, and had reaction studies been made with different H atom flux rates, more definite conclusions with respect to this question might have been obtained. Further clarification of this point could probably also have been derived from electron-spin resonance and matrix isolation investigations.

The finding that methane comprises 91% of the total hydrocarbon yield in the H atom–carbon reaction in this study and studies of other investigators (2, 9) is interesting in that different reaction conditions were used. In this study the initial H atom temperature before interacting with the carbon target was 2600°K whereas in the earlier studies hydrogen atom temperatures in the range 300°–373°K were used. Despite these differences the relative yield of methane with respect to the other hydrocarbons remained unchanged, implying that the factors responsible for the observed product distribution are independent of the H atom temperature. Surface-controlled phenomena appear to be dominant. The effective hydrogen atom pressure at the carbon target in this study was at least a factor of 10^3 lower than that used by other investigators (2, 9). If gas-phase reactions were important in determining the product distribution, a noticeable variation between the different investigations might be expected. That this was not the case again suggests that the carbon surface reactions largely control the product distribution.

Hydrogen Atom–Carbon Reaction as a Function of Temperature. The investigation of the H atom–carbon reaction as a function of temperature was followed by measuring hydrocarbon yields over the range 300°–1220°K. The best data were obtained for methane and showed that the reaction rate increased continuously with temperature over the entire range. Similar trends were found for ethane and propane. In two earlier studies a maximum in the reaction rate at about 720°K (9) and 820°K (8) was reported. The former study was based on the carbon removal rate after H atom attack while in the latter the methane yield was used as the rate indicator. These results are in disagreement with those obtained in this study. The maximum was justified in terms of the thermodynamic instability of methane at about 850°K (8, 9) at which temperature its free energy of formation changes from negative to positive. These arguments must be regarded as specious because the thermodynamic instability predicted for methane at 850°K requires H_2 and CH_4 to be at unit fugacity and for carbon to be in its standard state with a state of thermodynamic equilibrium existing among the three species. In the H atom–carbon reaction, these criteria for the application of thermodynamic reasoning do not apply, and, in fact, the conditions are deliberately chosen so that the results are kinetically controlled. If thermodynamic reasoning in any form could be applied to the reaction, and have any meaning, the system $4H + C = CH_4$ should have been considered. It is easily determined that for this process the reaction at 850°K has a free energy change of about −160 kcal. Therefore, a maximum in the H atom–carbon reaction rate cannot be justified on a thermodynamic basis. It is possible that in the earlier studies pyrolysis of the hydrocarbon reaction products on the high temperature carbon target may have been responsible for the observed maximum product-yield temperature. In the experimental design used in the present study pyrolysis of this type was kept to a minimum by virtue of the high pumping rate of the liquid helium cold finger.

It is also possible that the higher H atom temperatures used in this study compared with those previously reported may have resulted in the hydrocarbon-producing reactions occurring closer to the surface of the carbon target. If indeed this is the case, the opportunity for reaction product pyrolysis to occur during diffusion out of the pores in the carbon target would be reduced.

From the Arrhenius plot of the experimental data for methane shown in Figure 7, three fairly well defined different reaction regimes were found: 300-500°K, $E_a = 0.94 \pm 0.20$ kcal/mole; 500-1000°K, $E_a = 0.15 \pm 0.05$ kcal/mole; and 1000-1200°K, $E_a = 4.50 \pm 1.20$ kcal/mole. The data shown in Figure 7 must be considered as a pseudo-Arrhenius plot because the temperatures of the H atoms and the carbon target were considerably different whereas the simple reaction rate theory assumes equal temperatures for all reactants. Justification for three different reaction rate regions might be possible if data were available on the adsorption characteristics of H atoms on a carbon surface; however, this is not the case. Data on the adsorption characteristics of molecular hydrogen on some carbons (*13, 14, 15, 16*) are known. The data indicate that several different types of sites are available. In the most recent study (*16*), in the temperature range 80°-600°C a total of five discreet sites were characterized. It is possible that for atomic hydrogen several different types of reactive sites in carbon may exist, and this could possibly result in the different observed reaction rate regimes. As noted earlier, contrary to previous studies, a maximum in the reaction rate was not observed in this investigation at about 770°K. Instead at about 1000°K a sharp increase in rate was recorded. This is not easy to explain. A considerable body of evidence indicates that the basal plane edge atoms in carbon are more reactive than the non-edge basal plane atoms (*17, 18, 19, 20*). The latter sites are more predominant than the edge sites, and it may be that at temperatures above 1000°K the former sites become better able to participate in the reaction.

Although activation energies for the H atom–carbon reaction have been reported (*6, 9*), the values were based on the rate of carbon removal, not on the yield of methane as in this study. Ethane and propane are produced in significant quantities in addition to the methane, so the two sets of activation energies are not strictly comparable. However, in this study the yields of both ethane and propane followed similar trends with temperature as that of methane, and it is possible that the activation energies based on methane yields and carbon removal rates are not too different. King and Wise (6) have reported activation energies based on carbon removal rates of 9.2 and 7.1 kcal/mole in the range 365°-500°K, and 5.15 kcal/mole in the 450°-715°K region. These values are all considerably larger than those obtained in this study. Hydrogen atom temperatures below 370°K were used in the carbon removal-rate method of determining activation energies (6, 9) while in this study hydrogen atom temperatures of 2600°K were used. The latter have a thermal energy (translational) of about 5 kcal/mole whereas the former have about 0.7 kcal/mole. If the hydrogen atom translational energy is important energetically in the H atom–carbon reaction, the activation energies obtained in this study should be increased by about 4 kcal/mole to be comparable with those of the earlier studies. This improves somewhat the agreement between the two different sets of experimental activation energies.

Hydrogen Atom–Hydrocarbon Conversion Efficiency. The data obtained in the present study allow some limits to be placed on the efficiency with which hydrocarbons are produced in the H atom–carbon reaction. In the following calculations it is assumed that all hydrogen fed to the reactor leaves the effusion tube entirely as atoms. From data in Table III, calculations show that 1 mole of hydrogen atoms formed in the reactor results in the production of about 1.25×10^{-3} mole of CH_4, 1.2×10^{-4} mole of C_2H_6, and 7.5×10^{-6} mole of C_3H_8 at 30°C. If it is assumed that all hydrogen atoms effusing into the reactor can potentially react with the carbon target to produce hydrocarbons, then 0.6, 0.07, and 0.06% of the atoms are utilized to form CH_4, C_2H_6, and C_3H_8 respectively. These figures imply that 1 of every 140 hydrogen atoms entering the system is involved in the formation of hydrocarbons.

Of course, all of the hydrogen atoms formed in the reactor have an opportunity to interact with the carbon target. Using molecular beam properties and the geometry of the reactor system (21, 22, 23, 24), it can be shown that between 60 and 70% of all atoms leaving the effusion tube will interact directly with the carbon target. It is possible that some H atoms interact with the carbon target after undergoing reactor wall collisions. The number of such secondary target collisions is likely to be small because the target surface area is much smaller than the total internal surface area of the reactor. Assuming that only those H atoms interacting with the target directly after leaving the effusion tube are likely to produce hydrocarbons, the fraction of atomic hydrogen collision leading to CH_4, C_2H_6, and C_3H_8 is calculated at approximately 1.0, 0.1, and 0.1%, respectively. These numbers imply that about 1 of 83 H atoms colliding with the target produces hydrocarbons. At about 950°C, the hydrocarbon yield values are increased by about a factor of three, and hence the collision efficiency of H atoms to produce hydrocarbons becomes about 1 of 28 H atoms.

It is not possible to compare directly the H atom–hydrocarbon conversion efficiencies obtained in this study with those of other workers because either the data are not reported or the experimental conditions are not analogous to those used in the present study. In two studies (9, 11) where the latter condition holds, H atom conversion efficiencies were reported between one and two orders of magnitude lower than obtained in this investigation. It appears that the higher hydrogen pressure used in the earlier studies probably resulted in experimental conditions in which H atom recombination rates were substantially greater than in the present study with the corresponding dimunition in hydrocarbon-forming processes.

Conclusions

The reaction of atomic hydrogen with carbon at 30°C results in the production of 91% CH_4, 8.4% C_2H_6, and 0.6% C_3H_8. The formation of C_2H_4, C_3H_6, and C_4 species in the reaction is zero or close to zero. About 1.2% of the atomic hydrogen interacting with the carbon target is converted to hydrocarbons at 30°C. At 950°C this fraction increases to about 3.6%. Over the range 30°-950°C the hydrocarbon product distribution remains essentially unchanged. Previous reports of a maximum in the hydrocarbon yield at 720°-820°K were not substantiated. The previously reported maximum is believed to be a function of the experimental arrangements. The H atom–carbon reaction rate to produce methane has three distinct phases.

Literature Cited

1. Aramenko, L. J., *J. Phys. Chem. USSR* (1946) **20**, 1299.
2. Harris, G. M., Tickner, A. W., *Nature* (1947) **160**, 871.
3. Blackwood, J. D., McTaggart, F. K., *Australian J. Chem.* (1959) **12**, 533.
4. Shahin, M. M., *Nature* (1965) **195**, 992.
5. Vastola, F. J., Walker, P. L., Wightman, J. P., *Carbon* (1963) **1**, 11.
6. King, A. B., Wise, H., *J. Phys. Chem.* (1963) **67**, 1169.
7. Gill, P. S., Toomey, R. E., Moser, H. C., *Carbon* (1967) **5**, 43.
8. Coulon, M., Bonnetain, L., *Compt. Rend. Acad. Sci. Paris* (1969) **269**, 1469.
9. Wood, B. J., Wise, H., *J. Phys. Chem.* (1969) **73**, 1348.
10. Sanada, Y., Berkowitz, N., *Fuel* (1969) **48**, 375.

11. McCarroll, B., McKee, D. W., *Carbon* (1971) **9**, 301.
12. JANAF Thermochemical Tables, Thermal Research Laboratory, The Dow Chemical Co., Midland, Mich.
13. Barrer, M. R., Rideal, K. E., *Proc. Roy. Soc.* (1953) **A218**, 311.
14. Barrer, M. R., *J. Chem. Soc.* (1936) 1256.
15. Redmond, J. P., Walker, P. L., *J. Phys. Chem.* (1960) **64**, 1093.
16. Bansal, R. C., Vastola, F. J., Walker, P. L., *Carbon* (1971) **9**, 185.
17. Olander, D. R., Siekhaus, W., Jones, R., Schwarz, J. A., *J. Chem. Phys.* (1972) **57**, 408.
18. Olander, D. R., Jones, R. H., Schwarz, J. A., Siekhaus, W. J., *J. Chem. Phys.* (1972) **57**, 421.
19. Rosner, D. E., Strakey, J. P., *J. Phys. Chem.* (1973) **77**, 690.
20. Acharya, T. R., Olander, D. R., *Carbon* (1973) **11**, 7.
21. Davis, L., Nagle, D. E., Zacharias, J. R., *Phys. Rev.* (1949) **76**, 1068.
22. King, J. G., Zacharias, J. R., *Advan. Electron Phys.* (1956) **8**, 1.
23. Smoluchowski, M. V., *Ann. Physik.* (1910) **33**, 1559.
24. Knudsen, M., *Ann. Physik.* (1910) **31**, 633.

RECEIVED May 25, 1973. Work supported by Cleveland Consolidated Natural Gas Service Co. Inc.

6

Problems in Pulverized Coal and Char Combustion

DAVID GRAY, JOHN G. COGOLI, and ROBERT H. ESSENHIGH
Combustion Laboratory, The Pennsylvania State University,
University Park, Pa. 16802

Coals generate the greatest volatile matter yield if heated to reaction temperature at very high rates to prevent crosslinking reactions that may reduce yield. Dilute-phase instead of dense-phase reactions may also enhance yield by eliminating secondary capture of cracked volatiles. This view is supported by laboratory studies. In flames, the chars formed after pyrolysis burn at rates dominated by internal chemical reaction, not diffusion, with reaction in zone I or zone II. At higher temperatures ($\sim 2000°C$) reaction in zone I is evidently first order with low activation energy (6 kcal/mole). At lower temperatures for zone I, E = 40 kcal/mole. Zone II yields E = 20 kcal/mole with reaction order indeterminate but probably close to 0.5.

The dwindling supplies of natural gas and the predicted shortage of oil have initiated considerable research on the conversion of coal to easily used synthetic gas and oil. Many of these conversion processes yield a high proportion of by-product char whose reactivity has been questioned. For the overall process to be economically viable, the char must be recoverable as an energy source. The study of pulverized coal combustion is being reanalyzed to gain deeper understanding of several phases of the general process still not totally understood.

The combustion of pulverized coal particles can be broken down into two main processes: (a) evolution of volatiles (pyrolysis) and their subsequent combustion; and (b) heterogeneous combustion of the solid residue. Questions at issue in these two processes are: (a) What is the effect on pyrolysis yield of the rate of heating, which in pulverized coal systems can reach $10^{4}°C/sec$; and (b) what are the reaction order and activation energy of the subsequent char burnout? Answers to these questions can affect approach to design of both normal pulverized coal boilers and char burning boilers; the results can also influence the pyrolysis in the gasification process itself. In this article we address ourselves primarily to rate of heating effects on pyrolysis and mechanisms of reaction in char combustion.

Pyrolysis

Pyrolysis decomposition by heating is usually assumed to involve rearrangement of chemical bonds resulting in two or more new chemical products. When coal is pyrolyzed (in its simplest terms) the original organic components are split into a volatile fraction—the so-called volatile matter—and a solid residue, a matrix mainly of char—the so-called fixed carbon. Analysis shows that the volatile matter is really a mixture of compounds ranging from low molecular weight components (hydrogen, methane, carbon monoxide) to high molecular weight tars and bitumens. Interest in pyrolysis dates from the early 1800's when gas was first produced from coal but pyrolysis was also important in combustion and in gasification.

In coal pyrolysis the factor that most significantly affects the nature and proportions of the pyrolysis products is coal rank—*i.e.* (in general terms), the volatile matter content of a coal as measured under standard conditions. Beyond that, however, the products from even a single coal are very strongly affected by the pyrolysis temperature, rate of heating, particle diameter, and ambient atmosphere.

Heating Rate. Consider a coal sample being heated to a predetermined temperature to study pyrolysis behavior at that temperature. Such experiments are usually performed at heating rates of a few degrees per minute. As the sample heats, the changes are relatively minor (mostly loss of water and perhaps CO_2) until a temperature is reached at which decomposition effectively sets in. This decomposition temperature can be measured fairly precisely and is characteristic of a given coal. Clearly, as the coal sample is heated further to the required temperature, some decomposition will occur before this temperature is reached; the amount of this decomposition depends on the heating rate. If the heating rate is increased to reduce this prior decomposition, other factors become important. As the heating rate is significantly changed (for example, from degrees per minute to hundreds or thousands of degrees per second); at least four factors have to be considered:

(a) The decomposition mechanism is evidently a strong function of temperature; thus, the nature of the material being investigated at the study temperature depends on the rate of heating.

(b) As heating rate increases, the pyrolysis temperature likewise increases (*1*). At the usual low heating rates, pyrolysis of bituminous coals usually starts at 300°–400°C. At 1000°C/sec pyrolysis starts at 900°, and at 10,000°C/sec it starts at 1100°C.

(c) If particle size increases, diffusional effects are introduced. With small particles, the volatiles can be assumed to escape from the char matrix as rapidly as they are formed; thus, the overall rate of escape will be controlled by the chemical rate of volatiles production. Explosion studies by Ishihama (2) suggest that this is generally the case for particles less than about 50 μm. Particles above about 100 μm (but also dependent on heating rate) would seem to be large enough for volatile escape to be controlled by diffusion through the char matrix; Juntgen's experiments (1) further indicate that the diffusional escape for at least some of the components is also activated.

(d) With still larger particles (and for faster heating rates) the usual laboratory assumption that there are no temperature gradients through the particles is no longer valid. With nonisothermal conditions (that occur with particles greater than about 1 cm at the lower heating rates), a pyrolysis wave moves into the particle with unreacted material on the inside and pyrolyzing material on the outside. The rate of evolution is then a complex function of the rate of heat supply and the rate of progression of the pyrolysis wave. At very high rates of heat supply the distribution between heat for pyrolysis and heat for rise in matrix temperature moves in favor of the latter; high temperatures can be reached in very short times with small or negligible pyrolysis, even with significant temperature gradient through the particles. (With small particles there is no appreciable temperature gradient; this is the condition explained by Juntgen.)

Particle Size. The effect of particle size as a controlling factor in the rate of escape of volatiles can, in principle, be identified by experiment. If escape is influenced either by diffusional delay through the char matrix or by the rate of progression of a pyrolysis wave through the particle (another diffusional process), the rate of escape becomes a function of particle size. Conversely, if escape is controlled by chemical formation of the volatiles, the overall rate of change of volatile content (dV/dt) is particle-size independent. Ishihama's results (2) suggest that the transition size is about 100 μm. Direct support for this was obtained by Essenhigh (3) for single particles in the range 300 μm to 4 mm; his experiments showed a strong dependence of volatile burning time (equated to the pyrolysis time) on diameter. The relation was well approximated by $t_v = K_v d_o^2$; where t_v is the volatile burning or pyrolysis time; K_v is a constant of proportionality (about 100 cgs units), and d_o is the initial particle size in cm. A square law relation for the system was obtained theoretically (4), assuming a shrinking liquid-drop model of volatile loss from a carbon matrix.

With the large captive particles, the particles only pyrolyzed, with simultaneous combustion of volatiles and formation of a char particle. When pyrolysis stopped, char combustion started. This sequence of events is well known for large particles—particularly lumps on a grate— and it was also the accepted mechanism for small particle combustion of pulverized coal (about 1-100 μm). With very small particles, the rates of heating in a pulverized coal flame can exceed $10^4°C/sec$. At these heating rates Juntgen's experiments and analysis show that the pyrolysis temperature can be effectively increased to 1100° or 1200°C. Consistent with this, Howard and Essenhigh (5, 6, 7, 8) had previously reported that coal particles heating in a one-dimensional flame at 20,000°C/sec ignited (heterogeneously) at about 1000°C and started to pyrolyze at about 1200°C. The spacing between the heterogeneous ignition plane (determined by loss of O_2 and rise of CO_2, as well as visually) was physically displaced in the furnace by about 1 inch or 0.03 sec from the volatile loss plane. These conclusions were criticized by Kimber and Gray (9) but on grounds not pertinent to the argument. The original conclusions were not based at all, as inaccurately stated (9), on fixed carbon loss but on many other factors including the constancy of the volatile matter through the heterogeneous ignition/combustion region, separate locations of the heterogeneous and volatile flame fronts, and similar factors. Part of the basis for Kimber and Gray's criticism was neglect of the later reported "Q factor" phenomenon (described below); but introduction of a constant or variable Q factor into Howard's results changes only the magnitude not the logic of his argument. By contrast, unreported re-analysis of Csaba's data (10) on measured flame speeds in a cone are consistent with theoretical prediction (11) if prior pyrolysis at about 300°C is assumed with a heating rate of about 1000°C/sec or less. For particles of this size, however, the mechanism of pyrolysis is evidently not inconsistent with the assumption that pyrolysis is a volumetric reaction, uniform throughout the particle, independent of particle size, and controlled by the chemical rate of formation of volatiles. Howard, in particular, showed that extrapolating the square law escape equation down to pulverized coal size predicted pyrolysis times one or two orders of magnitude less than measured.

Swelling. The above discussion disregards additional well known aspects of behavior during pyrolysis, notably swelling and related problems. As early as 1910 it was reported that coal particles could form swollen "carbospheres" in flames. The earliest detailed investigation of the phenomenon appears to have been by Newall and Sinnatt (12, 13, 14) who also coined the phrase cenosphere (apparently renaming the carbosphere). The essence of their work, recently confirmed by Street et al. (15), was that the degree of swelling and the form of the resulting structure depended strongly on the maximum temperature and the nature of the ambient gas. Specifically, maximum swelling occurred at about 700°C, with true cenosphere formation (as originally defined) occurring only in neutral and reducing conditions—not in oxidizing conditions. Above 700°C the degree of swelling was reduced. Also true cenospheres in a flame should be indicative of neutral or reducing conditions. Microscopic examination of samples from flames can be misleading because the presence of large quantities of (for example) cenosphere-type particles may indicate the components that are not burning rather than the components that are burning. The particles that are burning may be present only in small sizes and quantities and will thus be overlooked. Other swelling measurements can be ambiguous or misleading. Comparison of crucible swelling numbers for a set of 10 coals ranging from non-caking to strongly swelling, with swelling in the same coals measured on single particles, showed much scatter among the single particles but an almost constant value of the average swelling factors (16).

The Q Factor. A recently established phenomenon of potential significance in pyrolysis and combustion deserves attention. This is the demonstration (9, 17, 18) that the volatile matter fraction can be greatly increased by significantly changing the experimental conditions. One factor common to the different experiments was a greatly increased rate of heating, and the investigators have tended to interpret the evident correlation as the causation. However there are difficulties with this interpretation. In the first place, rate of heating is generally not the only significant experimental change; to attain high heating rates it was generally necessary to reduce the particle density. The possibility of cracked volatiles being recaptured by adjacent char particles, if this is a plausible mechanism, is thereby reduced. Second, Badzioch and Hawksley showed that pyrolysis was delayed for 20 msec which corresponded to the mixing time between the hot and cold streams. At the estimated heating rates of $25,000°$ to $50,000°C/sec$, the conclusion of negligible pyrolysis during heating is consistent with Juntgen's predictions. The subsequent pyrolysis occurred, therefore, under isothermal conditions, and it is difficult to see how the prior heating rate history can then directly affect the subsequent isothermal pyrolysis.

Two Possible Mechanisms. We tentatively propose two causal mechanisms to account for the observed correlation of higher volatile matter loss with heating rate. Let us assume first that a coal sample is taken to reaction temperature infinitely fast (*i.e.*, exceeding about $10^4°C/sec$). The material then decomposing is essentially the raw, unaltered coal which is only partially ordered and therefore more prone to pyrolyze extensively than a more ordered structure. By comparison, coal heated slowly (less than $10^3°C/sec$) to reaction temperature reaches that temperature as a partially pyrolyzed but more ordered structure than the raw coal. The subsequent pyrolysis should therefore be less extensive.

A second explanation may also derive from the change in particle concentration. At particle concentrations usually encountered in volatile matter determinations, the upper levels of the sample can easily act as

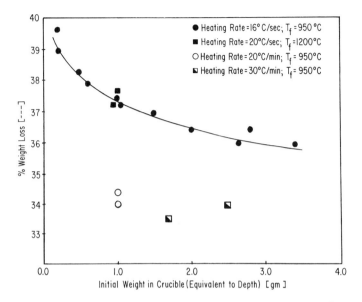

Figure 1. Per cent weight loss vs. initial weight of sample in crucible for different rates of heating and different final temperatures (T_f is final pyrolysis temperature achieved)

a trap for volatiles undergoing secondary pyrolysis with carbon deposition as they escape. Figure 1 illustrates a quick check of this suggestion showing that the per cent weight loss from a pyrolysis sample decreases with increasing sample weight (*i.e.*, depth). The increase of 4 percentage points in the weight loss as the total weight drops from 3 to 0.25 gram shows the importance of accurately following the ASTM standard if comparisons between coals are to be valid.

Capture. It is interesting to apply this capture concept to the experimental results of Badzioch and Hawksley. Their system was a vertical tube furnace through which coal particles were carried in a nitrogen stream under essentially isothermal conditions. (Through the heatup period, of about 20 msec, the extent of pyrolysis as mentioned above was negligible.) The particles were injected at the top of the furnace, collected, quenched, and the weight loss (ΔW) of pyrolyzed char was determined after isothermal reaction for varying times and at different temperatures. The volatile matter in the partial char products (VM') was then determined by standard methods and correlated with the per cent weight loss of the original coal (ΔW). The volatile per cent in the char (VM') was recalculated to yield a volatile loss from the original coal (ΔV), and an empirical correlation was obtained:

$$\Delta W = Q \times \Delta V + \Delta W_o \tag{1}$$

Here, Q is a (constant) multiplier factor representing the increase of volatile yield in the stated furnace condition over the standard volatile matter yield, and ΔW_o was interpreted by the investigators as an error intercept which, being small, was ignored. The multiplier, Q, was 1.3-1.5 for weakly swelling coals and 1.4-1.8 for highly swelling coals. Similar results were also obtained on similar equipment by Kimber and Gray (9).

These results help support our explanation of the observed correlations. Badzioch and Hawksley discuss their results essentially in terms of an increase of volatile matter yield by the conditions of experiment over the standard yield obtained by conventional test methods. However, we may invert our point of view and argue that their experimental conditions are the ones that give us the "true" or intrinsic volatile matter yield, and the crucible test provides abnormal experimental conditions that decrease the yield below the true value by reason (for example) of reordered structure during slow heating and/or volatile matter cracking and capture. Figure 1 is thus consistent with (but does not prove) the existence both of a capture effect and a variable effect with depth. Amplifying this capture postulate, let us assume the existence in crucible tests of a capture factor, α, such that

$$VM_o = VM_{act}(1-\alpha) \quad (2)$$

where VM_o is the volatile matter content of the original coal measured by conventional crucible tests, VM_{act} is the real or actual volatile matter content and both VM_o and VM_{act} are expressed as percentages of the original dry ash free (daf) coal. Here α represents the fraction of the true volatile matter content of the coal which is captured or retained during the proximate analysis. Similarly, the volatile matter of the resultant char, R, can be expressed as

$$R = R_{act}(1-\beta) \quad (3)$$

where R_{act} is the real volatile matter content of the char and β is the capture factor of the char. Here again R and R_{act} are expressed as percentages of the original daf coal. Because the char will, in general, have a different chemical composition and a different physical structure from the parent coal, β can be expected to be different from α.

The weight loss for a given devolatilization run can then be expressed as:

$$W = VM_{act} - R_{act} \quad (4)$$

$$\text{or } W = \frac{VM_o}{1-\alpha} - \frac{R}{1-\beta} \quad (5)$$

By the original authors' definition of ΔV, i.e., $\Delta V = VM_o - R$, we obtain:

$$\Delta W = \frac{\Delta V}{1-\alpha} + R\left(\frac{1}{1-\alpha} - \frac{1}{1-\beta}\right) \qquad (6)$$

and by substitution for R and rearranging,

$$\Delta W = \frac{\Delta V}{1-\beta} + VM_o\left(\frac{\alpha-\beta}{(1-\alpha)(1-\beta)}\right) \qquad (7)$$

which has the form of the empirical Equation 1 with a slope $\frac{1}{1-\beta}$ and an intercept $VM_o\left(\frac{\alpha-\beta}{(1-\alpha)(1-\beta)}\right)$.

In the nomenclature used previously we then have:

$$Q = \frac{1}{1-\beta} \qquad (8)$$

and

$$\Delta W_o = VM_o\left(\frac{\alpha-\beta}{(1-\alpha)(1-\beta)}\right) \qquad (9)$$

If α is less than β, then the intercept ΔW_o will be negative. Since β, by definition of its function, must be positive and less than 1, the slope will always be greater than or equal to 1. Thus if α is less than β, a non-zero value of volatile loss (ΔV) at zero weight loss (ΔW) is predicted as being

$$\Delta V\,(\Delta W = 0) = VM_o\frac{(\beta-\alpha)}{(1-\alpha)}$$

Physically, this situation could arise when a coal particle is heated at such low temperatures that no weight loss occurs during the usual devolatilization, but some slight changes in the chemical and physical structure of the intact particle do take place. If these changes are such that the capture factor of the char (β) is not equal to the factor for the coal (α), an apparent volatile loss would then be recorded for α less than β.

The above analysis accounts qualitatively for the difference between fast and slow heating. Data are insufficient for an independent qualitative comparison, but the data of Figure 1 may be indicative. Here the volatile matter yield increases when the sample weight decreases; if this is interpreted entirely as a Q factor, because of capture, the maximum increase possible is about 4 percentage points in 36%, giving a Q factor of 1.1. This is far from the value of 1.5-1.8 found by Badzioch and Hawksley. Unless the coal used for Figure 1 is anomalous, the difference suggests that two significant factors are operating; as suggested above, the second factor is greater ordering of the coal structure (or recrystallizing) during slow temperature rise which binds in material that would otherwise be lost by pyrolysis in a Badzioch and Hawksley type of experiment. Figure 1 also supports this view. The four data points obtained at heating rates of 20-30°C/min, as against 16-20°C/sec, show a further significant drop in volatile matter yield.

Heterogeneous Combustion

The reactions of carbon with gases have been studied intensively for over a century and the results have been reviewed extensively in the last decade (*19, 20, 21, 22, 23, 24*). Although the broad outline of the processes is substantially accepted, certain points are still disputed. In our view, the dominant reaction step in carbon oxidation at certain temperatures is still open to question. There are logical objections to the association of a high activation energy with a near unity reaction order, particularly if this is identified as a desorption process. At the same time there seems to be a contradiction between the conclusion reached by logical analysis and that reached from quantitative data; however, we propose that the higher probability of validity lies with the logical argument on the basis of present evidence.

In the reaction of a gas with a solid surface it is generally agreed that the process can be divided into four basic steps:

(a) mass transfer by diffusion of the gas reactant to the solid surface;

(b) chemisorption of the gas reactant on the solid surface;

(c) desorption of the gas product from the solid surface carrying with it one or more of the underlying atoms previously part of the solid; and

(d) mass transport by diffusion of the gaseous product away from the surface.

The steps additionally necessary to reach the solid surface may involve transport down pores within the solid into the interior; the reactant molecule may dissociate (if it can) as it approaches the solid surface; dissociated or undissociated, the molecules or atoms may adsorb immediately or may move over the surface before final capture (mobile adsorption); and when adsorbing, the ease of adsorption may or may not depend on the coverage (thus affecting the extent of adsorption and the adsorption isotherm to be obeyed).

Diffusion, Adsorption, Desorption. For our purposes here, to help us focus on the point of dispute mentioned above, we take a simple picture of diffusion, adsorption, and desorption at the exterior surface alone (no internal pore diffusion) and assume Langmuir kinetics. If the specific reaction rate (defined as rate of mass removal from the solid surface per unit area in unit time) is written as R_s, then for velocity constants of the diffusion, adsorption, and desorption, k_o, k_1, and k_2, respectively, R_s takes the quadratic form (25):

$$R_s^2 - (k_o p_o + k_2 + k_o k_2/k_1)R_s + (k_o p_o/k_2) = 0 \qquad (10)$$

where p_o is the mainstream oxygen partial pressure. For (a) fast diffusion and (b) fast desorption, Equation 10 (25) reduces to respectively, the Langmuir isotherm and to the "resistance" equation. (There is not, in fact, a total analog with Ohm's law; the analogy does not lead to Equation 10. The other specific case, of fast adsorption, leads to a factorizable expression showing that the controlling reaction mechanism can change discontinuously at a particular temperature.)

If dissociation, pore diffusion, etc. are now included, Equation 10 becomes more complex but remains essentially invariant from a topological point of view. The velocity constants can also be expanded, in particular with the diffusion velocity constant shown to be a function of Reynolds and Schmidt numbers; the other velocity constants include dependence on activation energies. Details of the supplementary equations are given below as needed or in the reviews previously quoted.

Reaction Kinetics. The central issue now is the different reaction order—and therefore mechanism—over different temperature ranges. Following Hottel and associates (26, 27, 28) there was a tendency to rely on the resistance equation and to identify the lower temperature behavior (roughly below 1000°C) with a chemical resistance and the higher temperature behavior with a diffusional resistance. The recognition of at least three resistances (even neglecting pore diffusion and dissociation effects, etc.) calls for a reexamination of the identifications. A further basis for reexamination is the numerical values involved in the different regions. For example, the low temperature chemical control or dominant region is identified with the first-order reaction in oxygen and an adsorption condition but with a high activation energy. However there are simple experiments quoted by Trapnell (29) whose logical consequences forbid a high activation energy for adsorption of oxygen on carbon. The controversy then centers on the two equivalent points: (a) If the dominant mechanism is adsorption, the reaction is first order (or approximately so as reported), but the activation energies must be low (less than 10 kcal/mole) in contradition to values reported; or (b) if the dominant mechanism is desorption, the activation energy is high as reported but the reaction order must be zero or one-half in pore diffusion, generally contradicting reported values. To resolve this point, we ask: how valid are the reported values?

Trapnell quotes actual values of adsorption activation energies of oxygen on diamond, showing values increasing with coverage (implying a non-Langmuir isotherm) from a low of 4.3 to a high of 23 kcal/mole. The more logical nature of this summarized evidence, however, depends less on the actual magnitudes of measurements and more on physical consequences:

(a) Charcoal will chemisorb substantial quantities of oxygen at room temperature in seconds which requires either a very high velocity constant or a very low activation energy;

(b) heats of adsorption measurements could be carried out at room temperature, again implying a fast rate of adsorption for the measurements to be possible; and

(c) most convincing of all, oxygen could be physically adsorbed at liquid air temperatures up to $-70°C$, but above $-70°C$ it was always chemisorbed, again consistent with the need for a low activation energy for adsorption.

Conversely, certain results are often quoted that appear to support a high value for the adsorption activation energy. Notably this is the case for the many investigations following Langmuir carried out at low pressure to eliminate boundary layer diffusion complications. These fall mostly into two groups: first, beds of particles or samples and second, fine filaments, usually electrically heated. However, as Blyholder and Eyring (30) have pointed out, in such systems the sample was heated but the gas was not. The gas–solid temperature difference then affects the results. One supported explanation postulates that the number of gas molecules able to absorb on the carbon surface is that whose energy exceeds some minimum E^*, and this number is relatively small when the gas is at ambient temperature. E^* falls with rising *sample* temperature giving rise to a relatively large temperature dependence for the reaction which is interpreted as a high adsorption activation energy. The argument that the gas molecules undergo sufficient solid surface collisions to reach the solid temperature is contrary to the assumption of an eliminated diffusion boundary layer; after a solid surface collision, the next collision will probably be with the container at ambient temperature. For filament experiments where the filament was heated electrically, there is the additional objection, first raised by Strickland-Constable (31), that the reaction is often severely modified by thermionic emission.

These two objections do not leave us with much confidence in the absolute numerical values quoted for the adsorption activation energy. When care was taken in low pressure experiments to preheat the gas, Blyholder and Eyring (30) reported differing values: in the temperature range $600°$-$800°C$ (at 15 μm Hg) the activation energy was 80 kcal/mole with a zero-order reaction which is consistent with desorption dominance of the overall reaction, and which changed with rising temperature in agreement with a Langmuir equation to 4 kcal/mole activation energy and a first-order reaction, consistent with adsorption dominance. Further, with thicker samples of the carbon-coated ceramic, pore diffusion effects were observed in the lower temperature range with the activation energy reduced to 40 kcal/mole and the reaction order rising to one-half (consistent with change from zone I to zone II). These latter results were also in good agreement with their re-analysis (30) of earlier data reported by Gulbransen and Andrew (32) with an activation energy of 37 kcal/mole and a reevaluated order of one-half. Gulbransen and Andrew described the reaction order as first, but this order plot did not go through the origin, whereas the subsequent replot against $p^{1/2}$ did. (However, the same data can be used to fit a Langmuir expression with, again, a good straight line for reciprocal reaction rate plotted against $1/p$.) However, their one totally unambiguous result was the decomposition study of a chemisorbed film in a vacuum (32) in which the only possible reaction was desorption and for which the activation energy obtained was 40 kcal/mole. This last experiment shows that the desorption activation energy is certainly in the range generally quoted for adsorption, and if both energies are truly comparable, the only basis left for distinguishing them is the reaction order.

Several investigators have reported first-order reactions in conjunction with activation energies between 20 and 40 kcal/mole. Careful examination, however, reveals the possibility of different interpretations. One of the earliest reports of identified first-order reaction was given by Parker and Hottel (26); however, Frank-Kamenetskii's reanalysis (33) shows a closer correspondence to one-half or two-thirds order reaction indicating, as later experiments would support, a zone II reaction with a true zero order. More recent investigators reporting a first-order reaction are likewise open to argument. Studies of anthracite- (34) and carbon smoke-generated (35) data, from which interpolated plots were obtained showing reaction rates rising linearly with oxygen concentration, are two examples. These cases have been discussed extensively elsewhere (36). Briefly, in the anthracite experiments the interpolation argument was possibly circular; in any event, the reported plots could not be regenerated from the raw data using a more general interpolation argument. In the carbon smoke experiments the reported plots showed non-zero intercepts as with the Gulbransen and Andrew experiments, and the original thesis (37) reported a better fit to a second order in oxygen concentration. This is a most unexpected finding, but it has been reported again by Magnussen (38). This result is outside all the existing theories.

We find that all data reported before about 1965 are arguable with respect to either activation energy or reaction order. Since then, four additional sets of data have appeared in two groups which are all more acceptable from the experimental definition point of view but in which the results from each group contradict each other. The first group covers experiments on small particles by Field (*39, 40*) and by Smith and Tyler (*41, 42, 43*); the second group covers larger single spheres by Froberg (*44*) and by Kurylko and Essenhigh (*45, 46*).

Activation Energy and Temperature. Field developed data that are respectably convincing in support of a first order in oxygen concentration (subject to the reliability of a diffusion boundary layer calculation. He did not obtain a definite activation energy; instead he reported that he could explain his results empirically by assuming a variable activation energy between 30 kcal/mole at 1300°K and 10 kcal/mole at 1800°K. This is consistent with our proposition that he was recording data in the transition range from a desorption to an adsorption mechanism, so we further support our proposition using his data.

According to Langmuir's equation with p_s as the surface oxygen concentration calculated as mentioned above from the main stream values, we have

$$R_s = \frac{k_1 k_2 p_s}{k_1 p_s + k_2} \tag{11}$$

Field presented his data in terms of (R_s/p_s) which he wrote as K_s. By expansion of the velocity constants and rearrangement we obtain:

$$\ln (R_s/p_s) = \ln k_1° - E_a/RT_s \tag{12}$$

where
$$E_a = E_1 + RT_s \ln [1 + b \cdot \exp(\Delta E/RT_s)] \tag{13}$$

with
$$b = (k_1°p_s/k_2°) \text{ and } \Delta E = E_2 - E_1$$

In these expressions we have three potentially variable parameters available for curve fitting; E_1, b, and ΔE. For convenience the curve fitting was performed against smoothed data using Field's method of smoothing in which he found he could write his effectively variable activation energy, E, as

$$E = RAT_s^2/(R_s/p_s) \tag{14}$$

We propose that the empirical E is equivalent to the more fundamentally derived E_a of Equation 13. Using calculated values of E as input data,

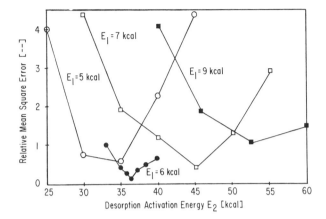

Figure 2. *Mean square error for* b = 0.000126 *and adsorption activation energy* E_1 = 5,6,7,9 *kcal vs. desorption activation energy* E_2. *Note that for* b = 0.000126 *minimum error occurs at* E_1 = 6 *kcal and* E_2 = 37 *kcal.*

Equation 13 was matched to the data by least-squares best fit. Figure 2 illustrates a typical set of the results from this process, and the best fit values obtained were:

E_1 = 6 kcal

E_2 = 37 kcal

b = 0.000126

$k_1°$ = 1.6 grams/cm²-sec-atm O_2

$k_2°/p_s$ = 1.27 × 10⁴ grams/cm²-sec-atm O_2

The backplot using these values is illustrated in Figure 3 which shows (R_s/p_s) as a function of T from Equation 11 compared with Field's original data. (Note: the optimizing was carried out on Field's activation energy Equation 15 which he obtained from his straight-line approximation. This is the source of the departure from the best fit at the lower temperatures.) Figure 3 also gives the curves for k_1 and for (k_2/p_s). There is an implication in the latter term that p_s is constant (but *see* additional discussion below).

These results show that at face value the agreement supports the proposition of a mechanism change from low temperature to high temperature. At the higher temperatures (above 1800°K), we agree with Field that the reaction must be first order or tending to first order. There is a point to argue or to discuss regarding the reaction order at the lower temperatures (below 1600°K). According to Equation 11, the true specific reaction rate, R_s, should be dominated by the desorption process which in standard theories is zero order so that $R_s \simeq k_2$. In Field's nomenclature this becomes $(R_s/p_s) \simeq (k_2/p_s)$, and we should have strictly the composite curve (solid line) splitting into a family of lines as T drops. Likewise we should expect increased scatter of the points as T drops. However a halving of the oxygen concentration should halve the value of (R_s/p_s)—if halving p_o (the mainstream value) also halves p_s—but the values of (R_s/p_s) at the lower temperatures are very small and subject to substantial error that can obscure any fine detail. The numerical tables given by Field clearly show that the effect of increasing the oxygen concentration at the same furnace temperature also increases the particle temperature substantially. This is a consequence that has been extensively discussed by Froberg (44) and Kurylko and Essenhigh (45, 46), and the increased particle temperature is primarily responsible for the increased reaction rate. Therefore at the lower temperatures, Field's data neither support clearly nor contradict our postulate of a zero-order reaction. The question is still open.

The equation that is the basis for the reanalysis (Equation 11) is normally applied to a nonporous surface. Application in this context can

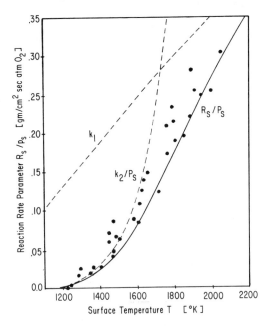

Figure 3. Calculated R_s/p_s *and its component rates* k_1 *and* k_2/p_s *and Field's data (39) vs. particle temperature*

be challenged to the extent that any mechanistic interpretation is questionable, but it clearly has excellent empirical value. There is still an argument for validity since partial penetration of oxygen into the carbon pores can be equated with an effective increase of the number of adsorption sites per unit area. Of course the oxygen partial pressure, p_s, then falls inside the pores but, for relatively small penetration where the reaction is proportional to p_s, the weighted contribution from the pores is near the mouth and the average p_s is not far below the surface value. At the other extreme of very low reactivity, the oxygen concentration can be quite high through the particle (lower end of zone I), and again p_s is an adequate approximation to the average. Between the extremes, theory shows the full effect (19), and then a true zero-order reaction becomes an apparent half-order reaction. This could be the condition at the lower temperatures, but it would mean that the activation energy would be only an apparent value that would be half the true value. Although this would support our argument for the inability to pick up the reaction order at the lower temperatures, we believe that the reaction is in zone I, and the velocity constant k in Equation 11 has the usual effective values for internal reaction that are referred to unit area of superficial surface.

In contrast to Field's data, Smith reports no change in activation energy with temperatures over about the same temperature range and with about the same range of values of (R_s/p_s). However the activation energy in most of the experiments was in the range of 20 ± 2 kcal/mole which implies to us that the reactions were truly in zone II. This conclusion is supported by later data (43) in which the utilization factor was calculated at mostly below or significantly below unity. An exception was the data set for 6 μm particles of semi-anthracite which was shown to be not incompatible with a higher activation energy. The difference between Field's data and Smith's data can be explained as attributable to a difference in pore size distribution since the chars from high rank coals have substantially different pore size distributions from those of low rank coals. The higher rank coal chars tend to be predominantly microporous while the lower rank chars tend to be highly macroporous (47). Smith assumes, following Field, that the reaction order in his experiments is first, but he presents no direct evidence.

Adsorption/Desorption. There is earlier evidence for the dominance of a zero-order desorption mechanism at low to moderate temperatures in pure carbon systems. Rosner and Allendorf (48) observed a change in activation energy of 31 kcal/mole at 1300°K to 0 at 1600°K and a change in order from 0.56 at 1200°K to unity at 1440°K. If pore diffusion were a factor, the change in oxygen reaction order would indicate a shift from desorption (true order of zero) to adsorption (true and apparent order of unity).

Some recent evidence also supports the adsorption/desorption theory. Hamor et al. (49) report an apparent order of one-half for a brown coal char under zone II conditions which indicates a true order of zero. Kimber and Gray (50) have burned 34 and 54 μm charcoal particles in oxygen over the temperature range 1400°-2800°K. For both particle sizes the activation energy went to zero at roughly 2400°K and became negative at higher temperatures. This zero and negative activation energy behavior could be a result of a chemically dominating adsorption rate with a low activation energy and the usual $T^{-1/2}$ dependence of the pre-exponential factor in the velocity constant. Kimber and Gray observed an order of roughly unity for their range of measurements.

Field, as mentioned, calculated the surface oxygen concentration using a diffusion calculation that also included a temperature correction. Taken at face value, the calculations and consequences for the reported reaction (even at the higher temperatures) are quite convincing. The calculations neglect, however, an effect first proposed by Froberg (44) and later investigated by Kurylko and Essenhigh (45, 46). Froberg found that rate data on 1/2-inch carbon spheres burning in air and oxygen depended only on sample temperature with no influence of oxygen concentration at all. From this he concluded that the reaction was zero order in zone I. (Additional work identified the regions of zones II and III in full agreement with theoretical expectations.) However, he also noted that the samples burned at a higher temperature in oxygen than they did in air which should be impossible for a truly zero-order reaction. To explain this anomoly, Froberg proposed that the source of the extra heat could be a change in the proportionation in the CO burnup inside and outside the sphere. If the primary (weight losing) reaction produces only CO as a zero-order reaction in zone I, the CO then can burn to CO_2 as it diffuses out of the sphere without altering the rate of the primary reaction because this is zero order and there is always surplus oxygen inside the sphere. The CO reaction however is relatively slow and when burning in air part will burn up in the boundary layer outside the sphere in air, and the heat from this reaction fraction will be lost directly from the reaction volume. In oxygen, however, a greater fraction of the CO can burn up inside the sphere and additional heat will go directly into the solid and subsequently will be lost to the furnace walls by radiation. The extra heat was estimated to be sufficient to raise the sphere temperature by 10°, 20°, or 30° compared with air, and this difference agreed with that observed. In Field's calculations an effect of this sort, with part of the heat of combustion not liberated near or at the particle surface, can affect the estimated particle temperatures appreciably and thereby affect all conclusions derived from the data.

Kurylko and Essenhigh (45, 46) subsequently examined this point in greater detail both experimentally and by extensive calculations and confirmed that Froberg's postulate was well based. During this work Kurylko also found that a change in location of the CO combustion region between inside and outside could generate both temperature and reaction rate oscillations.

Conclusions

From review of the pertinent literature, our conclusions regarding pyrolysis are:

(a) Rapid heating can raise coal samples to high temperatures without significant decomposition, and they can then pyrolyze at constant temperature with a yield of volatiles that is higher than can be obtained under any other experimental conditions.

(b) Pyrolysis at lower heating rates may promote part of the coal substance to crosslink during pyrolysis in the period of temperature rise to the final constant pyrolysis temperature. This crosslinking binds a material that would otherwise be able to escape as volatiles, thus reducing the volatile yield.

(c) If pyrolysis is also carried out in a dense packed bed of particles of some depth, as in the ASTM standard analysis for volatile matter, some fraction of the volatiles escaping from the lower sections of the sample may crack during escape and be captured in part by the upper fraction which acts as a trap.

(d) Analysis of trapping in a dense bed by introduction of a capture factor yields an expression relating weight loss in dilute phase after rapid heating to the loss determined by standard volatile matter methods, and this expression agrees with an empirical expression based on the experimental results obtained by Badzioch and Hawksley. This is believed to provide some support for the capture process. (A satisfactory analysis yielding a parallel result for crosslinking during heating is still incomplete.)

(e) The implications for interpretation of the results are that the difference between rapid and slow heating experiments are that rapid heating as such has no direct influence on the yield of volatiles; rather, the yield during isothermal pyrolysis following rapid heating to the reaction temperature should be recognized as the normal or true value, and the (lower) yield under other conditions should be regarded as anomalous value.

(f) The implications for application of the results are that coal reactions involving volatiles should be carried out following as rapid heating as possible to maintain a high volatile matter yield (that can be one and one-half as much of that measured in the ASTM analysis).

From review of the pertinent literature, our conclusions regarding the mechanisms of heterogeneous combustion of coal are:

(a) Boundary layer diffusion plays a minor to negligible role in the control of reaction rates of coal char particles during burnout.

(b) Best evidence is that the particles reacting in the flame are in the zone I or zone II control regions; that is, there is internal reaction with sometimes partial and sometimes total penetration of the reaction zone through the particle. The circumstances of partial penetration are of rather dense particles with relatively small pore diameters and/or particle sizes at the upper end of size scale. Total penetration can occur either with very small particles (6 μm identified in one instance) with relatively porous particles either from middle or low rank coals or with more dense particles that have opened up during combustion.

(c) The best evidence is that particles burning in zone I have reaction orders and activation energies that are in the region of zero order and 40 kcal, respectively, in the region of 1000°K, changing to first order and less than 10 kcal (reanalysis of existing data yielded 6 kcal) in the region of 1700°C.

(d) Particles burning in zone II have activation energies that remain reasonably constant over the range 1000 to 2000°C. The reaction order is unknown (or indeterminate) although theory would indicate 0.5.

(e) Common belief in the overall reaction order of unity above 800°C does not appear to be well founded. Much quoted data can be shown to be suspect or ambiguous. The fully definite experiments on reaction order and identification of the dominant reaction step at higher temperatures have yet to be performed.

Literature Cited

1. Juntgen, V. H., Van Heek, N. H., *Fuel* (1967) **47**, 103.
2. Ishihama, W., *Int. Conf. Director Safety Mine Res., 11th*, Warsaw, Poland, 1961.
3. Essenhigh, R. H., Ph.D. Thesis, Sheffield University, England, 1959.
4. Essenhigh, R. H., *J. Eng. Power* (1963) **85**, 183.
5. Howard, J. B., Essenhigh, R. H., *Combust. Flame* (1965) **9**, 337.
6. Howard, J. B., Essenhigh, R. H., *Combust. Flame* (1966) **10**, 92.
7. Howard, J. B., Essenhigh, R. H., *Ind. Eng. Chem., Process Des. Devel.* (1967) **6**, 74.
8. Howard, J. B., Essenhigh, R. H., *Symp. (Intern.) Combust., 11th*, Pittsburgh, p. 399, 1967.
9. Kimber, G. M., Gray, M. D., *Combust. Flame* (1967) **11**, 360.
10. Csaba, J., Ph.D. Thesis, Sheffield University, England, 1962.
11. Essenhigh, R. H., Csaba, J., *Symp. (Intern.) Combust., 9th*, New York, p. 111, 1963.
12. Newall, H. E., Sinnatt, F. S., *Fuel Sci. Pract.* (1924) **3**, 424.
13. Newall, H. E., Sinnatt, F. S., *Fuel Sci. Pract.* (1926) **5**, 335.
14. Newall, H. E., Sinnatt, F. S., *Fuel Sci. Pract.* (1929) **7**, 118.
15. Street, P. J., Weight, R. P., Lightman, P., *Fuel* (London) (1969) **48**, 343.
16. Essenhigh, R. H., Yorke, G. C., *Fuel* (1965) **44**, 177.
17. Loison, R., Chauvin, R., *Chim. Ind. Milan* (1966) **91**, 269.
18. Badzioch, S., Hawksley, P. G. W., *Ind. Eng. Chem., Process Des. Devel.* (1970) **9**, 521.
19. Walker, P. L., Jr., Rusinko, F., Austin, L. G., *Advan. Catal.* (1959) **11**, 133.
20. Thring, M. W., Essenhigh, R. H., in "Chemistry of Coal Utilization, Supplementary Volume," p. 754, John Wiley and Sons, New York, 1963.

21. von Fredersdorff, C. G., Elliott, M. A., in "Chemistry of Coal Utilization, Supplementary Volume," p. 892, John Wiley and Sons, New York, 1963.
22. Field, M. A., Gill, D. W., Morgan, B. B., Hawksley, P. G. W., "Combustion Pulverised Coal," BCURA, Leatherhead, 1967.
23. Essenhigh, R. H., *Univ. Sheffield Fuel Soc. J.* (1955) **6**, 15.
24. Mulcahy, M. F. R., Smith, I. W., *Rev. Pure Appl. Chem.* (1969) **19**, 81.
25. Essenhigh, R. H., Froberg, R., Howard, J. B., *Ind. Eng. Chem.* (1965) **57**, 33.
26. Parker, A. S., Hottel, H. C., *Ind. Eng. Chem.* (1936) **28**, 1334.
27. Hottel, H. C., Stewart, I. M., *Ind. Eng. Chem.* (1940) **32**, 719.
28. Tu, C. M., Davis, H., Hottel, H. C., *Ind. Eng. Chem.* (1934) **26**, 749.
29. Trapnell, B. M. W., "Chemisorption," Butterworth, London, 1955.
30. Blyholder, G., Eyring, H., *J. Phys. Chem.* (1957) **61**, 682.
31. Strickland-Constable, R. F., *Trans. Faraday Soc.* (1944) **40**, 333.
32. Gulbransen, E. A., Andrew, K. F., *Ind. Eng. Chem.* (1952) **44**, 1034, 1039, 1048.
33. Frank-Kamenetskii, D. A., "Diffusion and Heat Exchange in Chemical Kinetics," Princeton University Press, Princeton, 1955.
34. Thring, M. W., Beer, J. M., *Proc. Anthracite Conf. Mineral Ind. Exp. Sta. Bull. No.* **75**, Pennsylvania State University, p. 25, 1961.
35. Lee, K. B., Thring, M. W., Beer, J. M., *Combust. Flame* (1962) **6**, 137.
36. Essenhigh, R. H., Amer. Soc. Mech. Eng., Paper **70-WA/Fu-2** (1970).
37. Lee, K. B., Ph.D. Thesis, Sheffield University, England, 1961.
38. Magnussen, B. F., *Symp. (Intern.) Combust., 13th*, Pittsburgh, p. 869, 1970.
39. Field, M. A., *Combust. Flame* (1969) **13**, 237.
40. Field, M. A., *Combust. Flame* (1970) **14**, 237.
41. Smith, I. W., *Combust. Flame* (1971) **17**, 303.
42. Smith, I. W., *Combust. Flame* (1971) **17**, 421.
43. Smith, I. W., Tyler, R. J., *Fuel* (1972) **51**, 312.
44. Froberg, R. W., Ph.D. Thesis, Pennsylvania State University, 1967.
45. Kurylko, L., Ph.D. Thesis, Pennsylvania State University, 1969.
46. Kurylko, L., Essenhigh, R. H., *Symp. (Intern.) Combust., 14th*, Pittsburgh, p. 1375, 1972.
47. Gan, H., Nandi, S. P., Walker, P. L., Jr., *Fuel* (1972) **51**, 272.
48. Rosner, D. E., Allendorf, H. D., *Carbon* (1965) **3**, 153.
49. Hamor, R. J., Smith, I. W., Tyler, R. J., *Combust. Flame*, in press.
50. Kimber, G. M., Gray, M. D., *Conf. Ind. Carbon Graphite, 3rd*, p. 278, 1971.

RECEIVED May 25, 1973.

7

Coal Devolatilization in a Low Pressure, Low Residence Time Entrained Flow Reactor

R. L. COATES, C. L. CHEN, and B. J. POPE

Chemical Engineering Department, Brigham Young University, Provo, Utah 84601

> *The amount and composition of gaseous volatile matter evolved during extremely rapid pyrolysis of a bituminous coal were studied experimentally. Continuous, rapid devolatilization of 1–4 lbs of coal/hr was achieved at atmospheric pressure by mixing the finely ground coal, entrained in a stream of hydrogen or nitrogen, with hot gas from a hydrogen–oxygen combustor, thereby heating the gas as high as 2500°F within 0.01–0.3 sec. Rapid quenching was done with a water spray. Total volatiles greatly exceeded the ASTM volatility of the coal, and as much as 14% of the coal was converted to methane, ethylene, and acetylene.*

Studies in which finely ground coal was heated very rapidly have shown that the fraction of the coal that can be volatilized increases with both the rate of heating and the final temperature to which the coal is heated. For example, Eddinger et al. (1) have presented data from an entrained flow reactor which show that volatile products amounting to 49.9% of the coal fed may be produced from a finely ground coal having an ASTM volatility of only 35.5%, even though maximum reactor temperature was less than the 950°C reached in the standard volatility test. Kimber and Gray (2) reported coal pyrolysis data in an entrained flow reactor operated as high as 2200°K. They observed volatiles as much as 87% greater than that from the standard test, and they concluded that both higher heating rates and higher final temperatures increase the amount of volatile products. Another characteristic of high-rate, high-temperature pyrolysis of coal that is not found in normal carbonization is the production of significant quantities of acetylene and ethylene in the pyrolysis gas. These products are commonly observed during coal pyrolysis in a plasma or by flash heating (3, 4).

The present study was done to investigate rapid coal pyrolysis brought about through rapid mixing of finely ground coal with hot combustion gases. We were particularly interested in evaluating the potential of this procedure for increased yields of volatile matter and for the production of unsaturated hydrocarbons as constituents of the volatiles.

Experimental

An entrained flow reactor was designed in which the finely ground coal could be mixed rapidly with oxidizing combustion gases. The combustion gases came from a premixed flame of pure oxygen with hydrogen. The reactor volume was designed for short residence times, and the products were quenched by water spray immediately downstream of the reactor.

A diagram of the reactor is shown in Figure 1. The reaction tube was made of alumina. This tube was placed inside an annular electrical heating element for preheating and to reduce heat loss during the run.

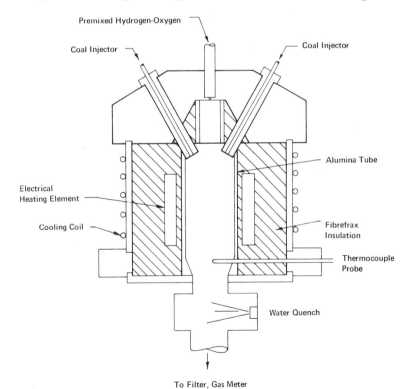

Figure 1. Schematic of reactor

Table I. Coal Analysis—Weight Per Cent as Received
(coal size, −200 mesh)

Proximate	Wt %	Ultimate	Wt %
Moisture	5.65	Carbon	70.05
Ash	6.20	Hydrogen	5.76
Volatile matter	34.35	Nitrogen	1.30
Fixed carbon	53.80	Sulfur	0.64
	100.00	Oxygen	10.40
		Moisture	5.65
		Ash	6.20
			100.00

The reaction tube and heating elements were insulated with fibrous alumina and encased with a water-cooled section of 6-inch aluminum pipe. Reaction tubes 4 5/8 inches long and 3/4-, 1 1/4-, and 2-inches id were tested. Smaller diameter tubes permitted testing at reduced residence times. The water-cooled injector head was aluminum. The coal was injected through two copper injectors located 180° apart and at an angle of 30° with the centerline of the reaction tube. The impingement point for these injectors was 3 inches below the orifice through which premixed combustion gases were fed to the reactor. A platinum/13% platinum–rhodium thermocouple was inserted near the base of the reactor to record the reactor temperature.

The coal tested was a high volatile B Utah coal from the Orangeville, Carbon County area. Typical proximate and ultimate analyses of coal from this area are listed in Table I. The coal was dried, ball-milled, and screened to −200 mesh for these tests. The moisture as used in the tests was less than 1%.

The coal was entrained into a stream of carrier gas, either hydrogen or nitrogen, with an auger-driven feeder. A variable-speed auger drive was used to obtain feed rates ranging from 0.5 to 5.0 lbs of coal/hr. Entraining gas flows of from 13 to 15 scfh were used in the 1/4-inch diameter feed line.

Table II. Range of Feed Rate Variables

Variable	Range
Coal feed rate, lbs/hr	0.7–4.1
Oxygen/coal ratio	0.3–1.6
Combustion gas equivalence ratio	0.4–1.1
Coal carrier gas, 13–15 scfh	N_2 or H_2

Reactor Operating Conditions	Range
Average temperature, °F	1200–2500
Average residence time, sec	0.012–0.343
Space time conversion, lbs/C gasified/ft^3-hr	13–408
Steam partial pressure (reactor exit), atm	0.125–0.255
Hydrogen partial pressure (reactor exit), atm	0.194–0.553

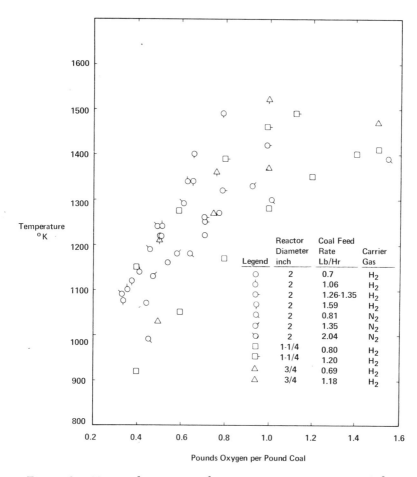

Figure 2. Measured reactor outlet temperatures vs. oxygen fed per pound of coal

The product gas was separated from the quench water, passed through a filter, and then passed through a gas meter. Samples of filtered gas were withdrawn for analysis by gas chromatography. The char was filtered from the quench water, dried, and analyzed for ash content to verify material balance calculations.

The operating parameters varied were the feed rates of the coal and combustion gases and the stoichiometry of the combustion gases. Run times following preheating of the reactor ranged from 2 to 22 min. The range of feed rate variables tested and the range of reactor operating conditions that resulted are in Table II.

Table III. Typical Data

	5-8-4	5-8-2
Reactor diameter, inches	2.000	2.000
Feed rates, lbs/hr		
Coal	1.590	1.590
Hydrogen carrier	0.082	0.082
Hydrogen combustion	0.069	0.102
Oxygen	0.540	0.800
Nitrogen carrier	—	—
Oxygen/coal ratio	0.339	0.503
Combustion equivalence ratio	1.022	1.025
Reactor temperature, °F	1445	1750
Volume gas produced (dry)		
Total, scfh	30.9	37.6
Carrier free basis, scf/lb coal	10.2	14.4
Gas analysis (dry, vol %)		
Hydrogen	74.12	69.88
Oxygen	0.20	0.42
Nitrogen	0.81	1.33
Methane	5.33	5.02
Carbon monoxide	15.35	18.41
Ethane	0.14	0.04
Ethylene	1.62	1.26
Carbon dioxide	1.55	2.06
Acetylene	0.88	1.58
Carrier free heating value, Btu/ft^3	449.4	430.2
Steam decomposed, %	18.28	28.79
Ash in char, %	9.3	14.9

Results

A total of 32 test runs were made with the 2-inch diameter reaction tube, 20 with hydrogen as the carrier gas, and 12 with nitrogen. Twelve tests were made with the 1 1/4-inch diameter reaction tube, and seven tests were made with the 3/4-inch diameter tube. Typical data obtained from these tests are in Table III.

Reactor Temperature. Analysis of the data showed that the primary variable governing the composition of the reactor products was the temperature. The temperature as indicated by the thermocouple measurements increased with the amount of combustion gas fed to the reactor per pound of coal. Although the reactor tube was electrically heated, from Gasification Tests

Run Numbers

5-8-1	6-9-4	7-31-1	8-23-1
2.000	2.000	1.250	0.750
1.590	2.043	1.670	1.180
0.082	—	0.082	0.082
0.140	0.167	0.167	0.113
1.050	1.260	1.340	0.900
—	1.020	—	—
0.680	0.616	0.802	0.762
1.066	1.060	0.997	1.008
1955	1966	2157	1913
45.9	56.5	60.3	39.7
19.6	21.2	27.3	21.1
69.75	26.38	64.23	67.14
0.15	8.30	1.65	1.39
0.45	46.89	4.12	4.66
3.85	1.81	2.34	3.22
20.85	12.96	23.48	18.83
0.01	0.00	0.00	0.01
0.64	0.12	0.19	0.52
2.51	2.68	2.39	2.57
1.79	0.86	1.60	1.66
395.0	358.9	365.7	392.1
38.81	48.67	46.86	36.84
15.8	16.5	15.68	13.94

the feed rates and heat transfer area were such that the heating elements exerted only a small effect on the reaction temperature, serving primarily to reduce heat losses. Figure 2 presents the measured temperatures as a function of the ratio of combustion oxygen per pound of coal; the effects of reaction tube diameter and coal feed rate are also shown.

Effect of Temperature. Figure 3 gives data showing the conversion of the carbon in the coal to the hydrocarbon gases methane, acetylene, and ethylene, and to carbon monoxide and carbon dioxide. The conversion data are plotted vs. the measured temperature without regard for variations in the other operating variables. These conversions were computed from the measured volume and composition of the gas produced, after condensation of the water vapor and the feed rate of the coal.

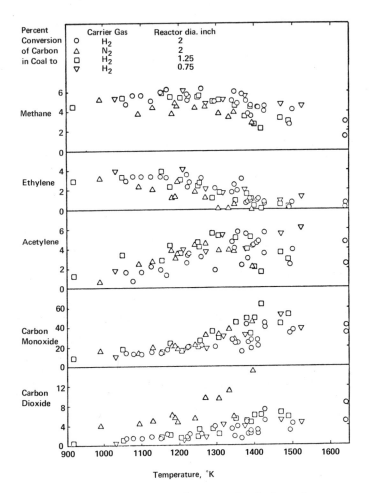

Figure 3. Conversion of carbon in coal to hydrocarbon gases, carbon monoxide, and carbon dioxide vs. reactor temperature

In contrast to gasification products of conventional low-pressure coal gasifiers, significant conversion to methane, acetylene, and ethylene was observed. The trends with reactor temperature are clearly evident. Methane conversion increases to a maximum and then decreases with increasing temperature, the conversion to acetylene increases with temperature, and the conversion to ethylene decreases with temperature.

The effect of replacing hydrogen as the coal carrier with nitrogen on the conversion to carbon oxides is also indicated in Figure 3. The conversion to carbon monoxide appears to depend principally on temperature; however the carbon dioxide yield is significantly greater with the lower hydrogen concentrations, resulting from the use of nitrogen carrier gas. Note that the amount of carbon dioxide produced relative to the amount of carbon monoxide is low compared with the products from conventional gasifiers.

Effect of Residence Time. The effect of average residence time in the reactor on the conversion to the three hydrocarbon gases is indicated by the data shown in Figures 4 and 5. Figure 4 shows conversion data from three different sizes of reactor tubes, each operated at a coal feed rate of 1.2 lbs of coal/hr. These data show only a slight effect of reactor size on the product yields. In Figure 5 the conversions for reactor temperatures in the range 1000–1300°K are plotted vs. the reactor space time,

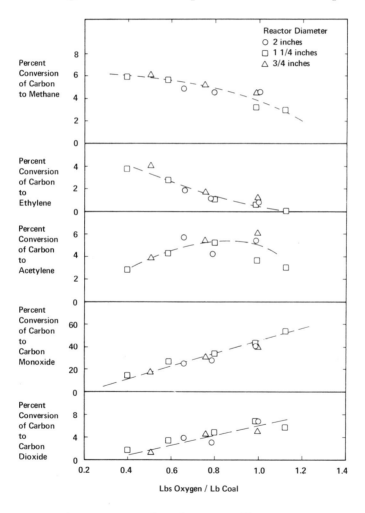

Figure 4. Conversion data showing small effect of varying reactor size. Data shown are for coal feed rate of 1.2 lbs/hr.

defined as the reactor volume divided by the computed volumetric flow rate at the reactor outlet conditions. The conversion to acetylene decreases as the space time is increased. However conversion to ethylene and methane increases with increasing space time; this variable has a more pronounced effect on the ethylene conversion than on the methane conversion. Another interesting observation from Figure 5 is that the devolatilization reactions producing the hydrocarbon gases are essentially completed in *ca.* 50 msec.

Effect of Hydrogen Concentration. Conversions to the hydrocarbon gases were generally higher the greater the concentration of hydrogen

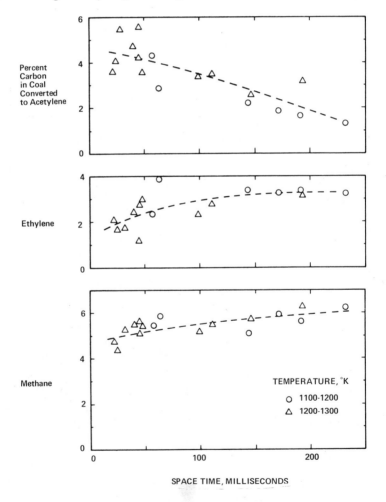

Figure 5. Conversion data showing small effect of average reactor residence time. Coal feed rate was 1.2 lbs/hr.

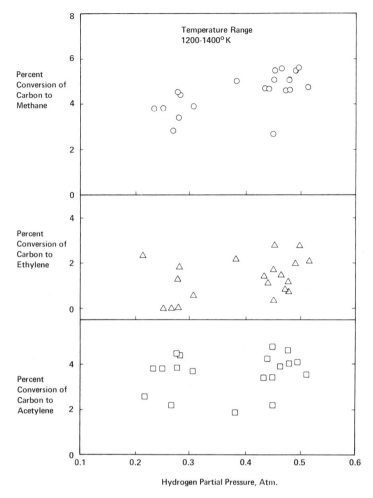

Figure 6. Data showing effect of hydrogen partial pressure on hydrocarbon yield

in the reactor. Data illustrating this effect are in Figure 6 where the conversions to methane, ethylene, and acetylene at 1200° to 1400°K are plotted vs. the hydrogen partial pressure at the reactor outlet. The conversion to methane is the most sensitive to this operating variable.

Although the observed effect of hydrogen concentration on the methane yield is in the direction expected from the hydrogenation reaction—*i.e.*, $C + 2H_2 = CH_4$—the equilibrium constant, K_p, for this reaction is much lower than the observed ratio of $P_{CH_4}/P^2_{H_2}$. The observed ratios are compared with the curve representing hydrogenation equilibrium in Figure 7. It seems clear from this comparison that the

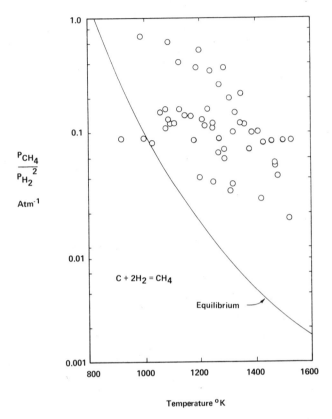

Figure 7. Comparison of equilibrium pressure ratios for the hydrogenation reaction with measured ratios

hydrocarbon gases are nonequilibrium species resulting from pyrolysis reactions.

Steam–Carbon Reaction. The composition and volumes of the product gas indicated that a significant fraction of the steam produced by the combustion gases reacted with the coal to form hydrogen and carbon monoxide. The calculated steam decomposition is plotted *vs.* the oxygen/coal ratio in Figure 8. This plot also shows the effect of the two carrier gases—hydrogen and nitrogen. The higher hydrogen concentrations resulting from the use of hydrogen carrier gas suppress the steam decomposition. The approach of the reaction $C + H_2O = CO + H_2$ toward equilibrium is indicated by the data presented in Figure 9. It is apparent from this comparison that the steam–carbon reaction is far from equilibrium for all of the run conditions tested.

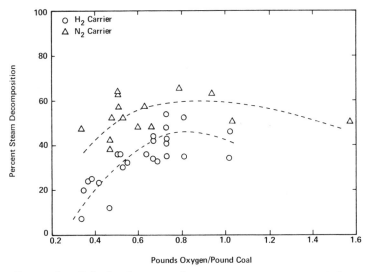

Figure 8. Calculated steam decomposition vs. oxygen fed per pound of coal

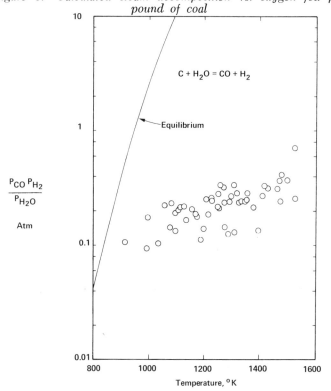

Figure 9. Comparison of equilibrium pressure ratios for the steam–carbon reaction with measured ratios

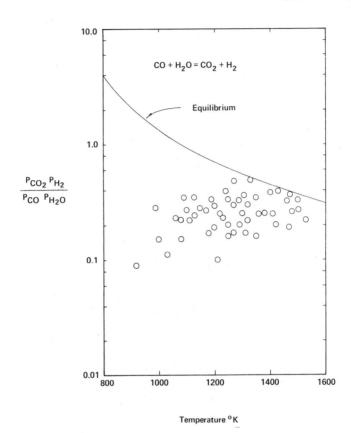

Figure 10. Comparison of equilibrium pressure ratios for the shift reaction with measured values

Shift Reaction. The conversion of carbon to carbon dioxide was rather low relative to the conversion to carbon monoxide, as mentioned above. In all cases the equilibrium constant, K_p, for the shift reaction, $CO + H_2O = CO_2 + H_2$, exceeded the observed ratio of $P_{CO_2}P_{H_2}/P_{CO}P_{H_2O}$ (Figure 10). The observed ratios approach the equilibrium K_p at the highest reactor temperatures.

Volume and Heating Value. The volume of dry gas produced less the volume of coal carrier gas fed to the reactor is shown in Figure 11 as a function of the oxygen/coal ratio. The volume produced increases uniformly with this ratio. The corresponding heating value of the dry, carrier-free gas is shown in Figure 12.

Combustion Equivalence Ratio. The effect of varying the equivalence ratio of combustion hydrogen to combustion oxygen was tested by operating the reactor with coal feed rate, oxygen-to-coal ratio, and carrier gas rate constant, and varying the combustion hydrogen feed rate.

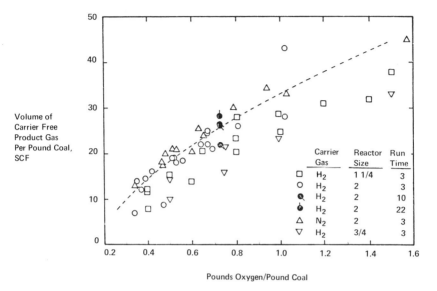

Figure 11. Net volume of product gas per pound of coal fed as a function of oxygen-to-coal ratio

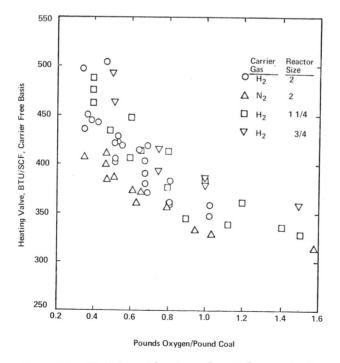

Figure 12. Variation of heating value of dry, carrier-free product gas with oxygen-coal ratio

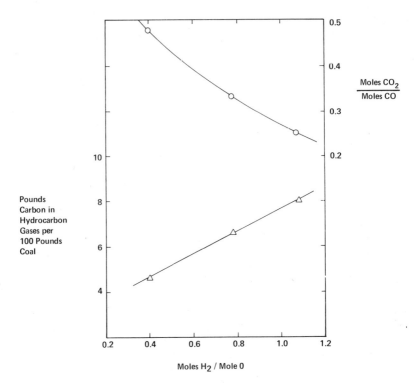

Figure 13. Effect of varying combustion gas equivalence ratio. Coal feed rate was 2 lbs/hr, and oxygen-to-coal ratio was 0.51.

Combustion hydrogen was varied to give an equivalence ratio (moles H_2 per mole O) from 0.4 to 1.1. As illustrated in Figure 13, the weight of carbon in hydrocarbon gases per 100 lbs of coal increased, and the molar ratio of carbon dioxide to carbon monoxide decreased with the equivalence ratio.

Summary

The observations and conclusions drawn from the experimental runs are summarized as follows:

(1) Principally as the result of extremely rapid heating, as much as 57% of a coal having ASTM volatile matter of 34% was converted to gaseous products. The overall yield of volatile matter depended primarily on the temperature of the reactor which in turn depended on the pounds of combustion gas fed per pound of coal. Variations in reactor space time caused only minor changes in the overall gas yield; high yields were achieved at space times as low as 0.012 sec. Space time conversions as high as 408 lbs of carbon/ft^3 of reactor volume/hr were achieved.

(2) Significant yields of methane, ethylene, and acetylene were produced. Up to 14% of the coal carbon was converted to these gases. The yield of ethylene decreased with increasing reactor temperature while the yield of acetylene increased. The conversion to methane passed through a maximum at a reactor temperature of about 1200°K. Maximum yields of methane, ethylene, and acetylene were 6, 4, and 6% of the coal carbon, respectively.

(3) The yield of hydrocarbon gases increased slightly with increasing hydrogen concentration. Residence times less than 50 msec are indicated for optimum yield of hydrocarbon gases. Comparison of the methane yield data with hydrogenation equilibrium indicates that the hydrocarbons in the product result not from hydrogenation reactions but from nonequilibrium pyrolysis reactions.

(4) Although a significant amount of steam decomposition was observed, conversion of carbon to carbon monoxide was substantially less than that predicted for the steam–carbon equilibrium. The conversion of carbon to carbon dioxide was much lower than predicted by the shift reaction equilibrium.

(5) The yield of hydrocarbon gases and the yield of carbon dioxide relative to carbon monoxide depends strongly on the stoichiometry of the combustion gas. Over-oxidized combustion gases cause the hydrocarbon yield to be reduced and the ratio of CO_2 to CO to be increased from the yields with stoichiometric combustion gases.

(6) The volume of gas produced per pound of coal increases uniformly with the oxygen/coal ratio. However, this increase may be simply the result of increased reactor temperature. At the ratio corresponding to maximum hydrocarbon gas yield the volume produced is 22 scf/lb of coal.

(7) The carrier-free heating value of the product gas decreases uniformly with increasing oxygen/coal ratio. At the ratio corresponding to maximum hydrocarbon gas yield, the heating value is in the range of 370–420 Btu/scf.

Literature Cited

1. Eddinger, R. T., Friedman, L. D., Rau, E., "Devolatilization of Coal in a Transport Reactor," *Fuel* (1966) **45**, 245.
2. Kimber, G. M., Gray, M. D., "Rapid Devolatilization of Small Coal Particles," *Combust. Flame* (1967) **11**, 360.
3. Bond, R. L., Ladner, W. R., McConnell, G. I., "Reactions of Coal in a Plasma Jet," *Fuel* (1966) **45**, 381.
4. Karn, F. S., Friedel, R. A., Sharkey, A. G., Jr., "Study of the Solid and Gaseous Products from Laser Pyrolysis of Coal," *Fuel* (1972) **51**, 113.

RECEIVED May 25, 1973. Work performed under contract to Bituminous Coal Research, Inc., with funds supplied by U.S. Office of Coal Research.

8

Pressurized Hydrogasification of Raw Coal In a Dilute-Phase Reactor

HERMAN F. FELDMANN, JOSEPH A. MIMA, and PAUL M. YAVORSKY

Pittsburgh Energy Research Center, Bureau of Mines, U. S. Department of the Interior, 4800 Forbes Ave., Pittsburgh, Pa. 15213

> *Raw, high volatile bituminous coal and lignite were hydrogasified in a continuous free-fall dilute-phase reactor using both hydrogen and hydrogen-methane mixtures. Reactor wall temperatures were varied from 480° to 900°C and reactor pressure from 500 to 2000 psig; most operations were at 1000 psig which is the most attractive operating pressure for a pipeline gas plant. These experiments established that this approach allows a high Btu gas production with 95% of the methane formed directly by the reaction of hydrogen with the coal and only 5% by methanation. In addition, this processing approach eliminates coal agglomeration problems encountered in other reactor systems without the necessity of coal pretreating.*

Raw coal can be converted directly to methane by allowing it to react with hydrogen. This approach is the basis of the Bureau of Mines Hydrane process (1), and the high thermal efficiency resulting from this direct process approach offers substantial potential economic advantages over other methods of producing pipeline gas (2). The basis of this process was first reported by Dent et al. (3). The thermodynamic advantages of producing methane directly—rather than by first converting the coal to synthesis gas which is then converted to methane after water-gas shift and methanation—were quickly recognized by U. S. investigators. For example, Channabasappa and Linden (4) concluded that hydrogenating coal to methane with hydrogen which is produced by steam–oxygen gasification of carbon is more thermally efficient than steam–oxygen gasification followed by methanation. However the experimental difficulties in directly hydrogenating raw coal to methane proved to be extreme because of the severe agglomerating properties of most American coals in high temperature, high pressure hydrogen atmospheres. This agglomeration problem caused a shift away from attempting to directly hydrogasify raw coal; instead, the coal was pretreated with air or oxygen to destroy its agglomerating properties. While the mild oxidation with air or oxygen was successful in preventing the coal from agglomerating, it adversely reduced its reactivity for methane formation. In fact, the reactivity of the pretreated coal is so reduced that it is impossible to produce, by direct hydrogenation of pretreated coal, a gas that has a sufficiently high concentration of methane to allow its introduction into a pipeline without costly physical separation of the hydrogen–methane mixture. Thus while the thermodynamic and chemical advantages of direct hydrogasification of raw coal were clear, the practical difficulties encountered in developing reactor systems to use raw coal required that the coal be pretreated before being contacted with hydrogen, and this reduced the process efficiency. Results of directly hydrogasifying pretreated coals in continuous reactors were reported by Institute of Gas Technology investigators (5, 6, 7).

The problem of processing agglomerating raw coal was solved at the Bureau of Mines when a technique was developed for directly hydrogenating raw coal in a free-fall, dilute-phase (FDP) reactor described by Hiteshue (8). Some results of FDP reactor experiments using raw coal (9, 10) were obtained for rather high pressures—1500 and 3000 psig. Even though operation with raw bituminous coal at 3000 psig does allow the direct production of raw product gases containing over 80% methane and carbon monoxide as low as 0.1 vol %, design considerations indicate pressures of about 1000 psig are more economical. This paper therefore summarizes our FDP reactor data at about 1000 psig. These data are useful for the design of the FDP section of the Hydrane process or other processes using similar conditions.

Experimental

Details of the experimental reactor system and its method of operation are given elsewhere (10). Briefly, the FDP reactor is a 3-inch id heated tube contained in a 10-inch pressure vessel. The coal is injected into the top of the 3-inch reactor through a water-cooled nozzle. The coal falls freely through the reactor tube concurrent with the reacting gas which is also injected at the top of the reactor. Because of rapid heating and a dilute solids phase, agglomeration is avoided; particles are plastic and sticky for only a short time during which particle–particle collisions are few. The heated length of the reactor for all except two of the experiments presented here is 5 ft. The residence time of the coal in the reacting zone is the reactor length divided by the average terminal velocity of the coal particles. The char produced in the FDP reactor is collected in a cooled hopper and analyzed after a run. Gas flow rates and compositions are measured over the steady-state portion of the run. Ordinarily the capacity of the pressurized char collector allowed a run of about 1 hr with approximately 50 min of steady-state operations during which data could be collected.

Results

Results of our most recent FDP reactor operations are summarized in Table I and the analyses of the feed coals used are listed in Table II. The main objectives of these experiments were: (1) to establish the feasibility of directly producing a high Btu gas by hydrogasifying raw coal in a continuous reactor at economical pressures, (2) to measure the yields and distribution of coal hydrogasification reaction products, and (3) to provide data for scaling up the FDP reactor.

Table I. Operating Data for FDP Hydrogasification of Raw Coal (Feed Coal is 50 × 100 Mesh Except Where Noted)

Parameter	Test No., IHR–					
	146	147	149	151	153	154
Temperature, °C	900	900	900	900	900	900
Pressure, psig	1000	1000	1200	1100	1100	2000
Coal	hvab	hvab	hvab	hvab	hvab	hvab
Coal rate, lb/hr	12.17	12.44	12.38	11.88	10.92	12.51
Feed gas rate, scfh	153.5	155.2	161.0	158.6	150.6	164.5
Hydrogen, vol %	50.5	56.0	53.0	48.0	99.2	52.5
Methane, vol %	41.9	42.3	44.5	49.2	0.2	46.4
Nitrogen, vol %	6.3	1.8	2.5	2.6	0.6	1.0
Total scf/lb	12.61	11.99	13.00	13.35	13.82	13.15
Hydrogen, scfh	77.5	86.9	85.3	76.1	149.4	86.3
Hydrogen, scf/lb	6.37	6.72	6.39	6.41	13.71	6.90
Product gas, scfh	169.6	171.8	175.5	167.1	143.8	175.0
Hydrogen, vol %	22.7	25.5	23.6	22.1	49.0	19.8
Methane, vol %	66.4	67.8	69.7	71.7	46.5	75.9
Ethane, vol %	0.3	0.4	0.3	0.1	0.3	0.1
Carbon monoxide, vol %	2.7	3.2	3.0	2.5	3.4	2.2
Carbon dioxide, vol %	1.0	1.1	0.8	0.5	0.2	0.4
Nitrogen, vol %	6.4	1.7	2.3	2.7	0.5	1.5
Product yield, Methane	3.97	3.93	4.09	3.52	6.10	4.52
Ethane, scf/lb	0.04	0.05	0.04	0.01	0.04	0.01
CO, scf/lb	0.38	0.43	0.43	0.35	0.45	0.31
CO_2, scf/lb	0.00	0.15	0.11	0.04	0.03	0.06
Feed H_2 reacted, scf/lb	3.21	3.33	3.55	2.46	7.23	4.13
Char residue, lb/lb	0.697	0.702	0.698	0.698	0.663	0.694
Condensed liquid, lb/lb						
Water	0.051	0.037	0.033	0.029	0.032	0.038
Oil	0.013	0.009	0.014	0.008	0.005	0.008
Residue moisture, lb/lb	0.009	0.012	0.011	0.008	0.003	0.018
Conversion, maf coal	32.5	32.5	33.1	32.8	36.5	33.0
Carbon, wt %	25.6	25.0	25.5	25.3	28.5	25.1
Hydrogen, wt %	64.2	66.6	66.4	65.2	70.0	66.6

Table I. Continued

Parameter	\multicolumn{6}{c}{Test No., IHR–}					
	146	147	149	151	153	154
Conversion, maf coal (continued)						
Sulfur, wt %	48.8	44.3	43.9	46.5	55.9	46.8
Nitrogen, wt %	25.1	24.6	27.2	30.6	38.0	26.4
Recovery, overall	96.3	96.0	96.3	93.2	95.7	96.2
Carbon, wt %	94.6	96.3	96.1	94.4	99.0	97.2
Hydrogen, wt %	98.9	95.2	97.1	92.4	94.5	97.3
Ash, wt %	100.2	100.2	104.0	99.4	103.4	92.4
	156	157	158	160	165	176[a]
Temperature, °C	850	850	900	900	850	850
Pressure, psig	1000	2000	2000	1500	1500	1000
Coal	hvab					hvab
Coal rate, lb/hr	12.84	13.00	12.61	12.29	12.94	12.47
Feed gas rate, scfh	157.8	161.7	160.9	161.3	158.8	156.2
Hydrogen, vol %	49.0	49.9	51.8	53.8	51.3	48.0
Methane, vol %	49.4	48.4	46.6	43.4	47.0	49.4
Nitrogen, vol %	1.6	1.7	1.6	2.6	1.7	2.2
Total scf/lb	12.29	12.44	12.76	13.12	12.27	12.53
Hydrogen, scfh	77.3	80.7	82.9	86.8	81.5	75.0
Hydrogen, scf/lb	6.02	6.21	6.61	7.06	6.30	6.24
Product gas, scfh	171.8	180.8	177.0	166.4	173.8	173.1
Hydrogen, vol %	22.4	18.1	18.0	19.7	21.7	22.8
Methane, vol %	71.4	79.0	78.7	75.2	73.4	71.6
Ethane, vol %	0.5	0.1	0.1	0.3	0.1	0.2
Carbon monoxide, vol %	3.2	0.5	1.1	1.4	2.1	2.5
Carbon dioxide, vol %	0.7	0.4	0.4	0.8	0.6	0.7
Nitrogen, vol %	1.4	1.7	1.6	2.4	2.0	1.9
Product yield, Methane	3.48	4.97	5.10	4.49	4.13	3.70
Ethane, scf/lb	0.06	0.01	0.01	0.04	0.01	0.03
CO, scf/lb	0.43	0.07	0.15	0.19	0.28	0.35
CO_2, scf/lb	0.09	0.06	0.06	0.08	0.08	0.10
Feed H_2 reacted, scf/lb	3.02	3.69	4.08	4.39	3.38	2.85
Char residue, lb/lb	0.700	0.658	0.703	0.697	0.696	0.706
Condensed liquid, lb/lb						
Water	0.036	0.042	0.050	0.049	0.058	0.048
Oil	0.018	0.015	0.004	0.005	0.012	0.022
Residue moisture, lb/lb	0.015	0.023	0.014	0.019	0.017	0.008
Conversion, maf coal	32.8	37.3	32.2	33.7	33.1	32.2
Carbon, wt %	25.0	30.0	25.0	24.2	23.3	24.0
Hydrogen, wt %	62.4	65.9	68.0	66.0	62.4	63.5
Sulfur, wt %	30.0	40.2	45.4	49.2	46.4	45.9
Nitrogen, wt %	21.3	34.2	29.7	30.6	25.8	24.7
Recovery, overall	95.1	94.1	98.1	95.8	96.8	96.9
Carbon, wt %	95.1	94.1	97.9	96.7	97.6	98.5
Hydrogen, wt %	95.4	102.0	100.6	96.9	100.9	98.5
Ash, wt %	106.8	102.2	104.3	101.8	96.9	98.0
	166	167	172	173	174	177
Temperature, °C	850	800	850	900[b]	850[b]	850
Pressure, psig	1200	1000	2000	1000	1000	1000

Table I. Continued

Parameter	\multicolumn{6}{c}{Test No., IHR–}					
	146	147	149	151	153	154
Coal	hvab					hvab
Coal rate, lb/hr	12.68	13.21	12.40	12.61	12.86	11.70
Feed gas rate, scfh	156.3	153.1	163.4	149.1	156.8	155.3
Hydrogen, vol %	49.2	48.4	50.6	52.7	49.6	99.3
Methane, vol %	48.7	48.2	46.4	45.2	47.0	0.4
Nitrogen, vol %	2.1	3.3	3.0	2.1	3.3	0.3
Total scf/lb	12.33	11.60	13.18	11.82	12.19	13.18
Hydrogen, scfh	76.9	74.1	82.7	78.6	77.8	154.2
Hydrogen, scf/lb	6.07	5.62	6.11	6.23	6.05	13.11
Product gas, scfh	168.7	170.3	171.9	163.0	174.1	150.9
Hydrogen, vol %	22.7	27.3	20.4	25.0	27.9	52.9
Methane, vol %	72.2	67.0	75.8	70.4	66.4	43.6
Ethane, vol %	0.1	1.0	0.1	0.2	0.8	0.2
Carbon monoxide, vol %	2.3	1.2	0.7	1.6	1.4	2.1
Carbon dioxide, vol %	0.5	0.2	0.2	0.4	0.6	0.3
Nitrogen, vol %	2.3	3.1	2.6	2.1	2.6	0.5
Product yield, Methane	3.60	3.05	4.39	3.76	3.26	5.57
Ethane, scf/lb	0.01	0.13	0.01	0.03	0.11	0.03
CO, scf/lb	0.31	0.15	0.10	0.21	0.19	0.27
CO_2, scf/lb	0.07	0.03	0.03	0.05	0.08	0.04
Feed H_2 reacted, scf/lb	3.04	2.10	3.84	3.00	2.27	7.21
Char residue, lb/lb	0.691	0.709	0.674	0.721	0.692	0.646
Condensed liquid, lb/lb						
Water	0.049	0.041	0.053	0.037	0.041	0.042
Oil	0.018	0.029	0.012	0.029	0.030	0.018
Residue moisture, lb/lb	0.019	0.014	0.015	0.018	0.019	0.014
Conversion, maf coal	34.8	30.5	35.0	31.4	33.7	38.2
Carbon, wt %	25.6	25.0	28.0	21.4	23.4	30.8
Hydrogen, wt %	63.5	59.1	64.3	62.4	59.1	67.1
Sulfur, wt %	42.8	52.7	44.9	61.8	42.1	45.9
Nitrogen, wt %	21.7	20.3	32.6	23.7	21.5	27.8
Recovery, overall	94.5	94.1	93.6	96.0	95.0	93.8
Carbon, wt %	92.9	94.0	93.4	99.1	96.9	95.5
Hydrogen, wt %	97.3	98.9	99.3	98.5	100.8	97.2
Ash, wt %	110.1	100.7	104.1	99.4	98.0	99.2
	178	180	181[c]	182	183[c]	186
Temperature, °C	800	900	900	900	850	900
Pressure, psig	1000	1000	1000	1000	1000	500
Coal	hvab					hvab
Coal rate, lb/hr	11.70	12.53	12.72	24.10	3.94	6.73
Feed gas rate, scfh	161.1	441.2	168.0	323.1	151.4	70.0
Hydrogen, vol %	98.9	99.3	99.2	52.6	51.6	99.0
Methane, vol %	0.4	0.5	0.5	45.7	46.5	0.4
Nitrogen, vol %	0.4	0.2	0.3	1.7	1.9	0.6
Total scf/lb	13.77	35.21	13.20	13.41	38.43	10.40
Hydrogen, scfh	159.0	438.1	166.7	170.0	78.1	69.3
Hydrogen, scf/lb	13.59	34.96	13.11	7.05	19.83	10.30

Table I. Continued

Parameter	Test No., IHR-					
	146	147	149	151	153	154
Product gas, scfh	147.5	412.6	155.1	335.0	152.1	76.8
Hydrogen, vol %	73.0	76.3	44.2	28.5	39.8	50.9
Methane, vol %	23.9	21.9	50.2	66.9	57.3	44.1
Ethane, vol %	0.7	trace	0.0	0.2	trace	0.0
Carbon monoxide, vol %	1.5	1.4	4.7	2.1	1.0	4.2
Carbon dioxide, vol %	0.2	0.0	0.2	0.3	0.1	0.0
Nitrogen, vol %	0.4	0.2	0.4	1.7	1.7	0.5
Product yield, Methane	2.96	7.04	6.06	3.17	4.25	5.12
Ethane, scf/lb	0.09	trace	0.00	0.03	0.0	0.0
CO, scf/lb	0.19	0.46	0.57	0.29	0.38	0.47
CO_2, scf/lb	0.00	0.00	0.04	0.04	0.04	0.0
Feed H_2 reacted, scf/lb	4.38	9.84	7.71	3.09	4.46	4.59
Char residue, lb/lb	0.705	0.648	0.630	0.708	0.602	0.618
Condensed liquid, lb/lb						
Water	0.039	0.028	0.042	0.038	0.004	0.036
Oil	0.029	0.010	0.007	0.019	0.054	0.017
Residue moisture, lb/lb	0.011	0.013	0.015	0.014	0.010	0.006
Conversion, maf coal	34.7	37.7	40.5	32.8	43.0	40.5
Carbon, wt %	28.1	31.6	33.2	26.0	36.2	33.4
Hydrogen, wt %	61.0	71.3	71.8	61.5	71.2	74.4
Sulfur, wt %	48.4	52.5	56.7	49.1	61.6	64.0
Nitrogen, wt %	21.9	38.1	41.6	25.3	43.7	46.0
Recovery, overall	81.8	96.6	95.9	92.6	91.9	90.6
Carbon, wt %	88.8	98.6	95.2	92.5	90.2	91.0
Hydrogen, wt %	86.3	99.6	92.9	94.3	97.2	95.5
Ash, wt %	112.6	97.7	99.4	107.7	101.8	99.6

Parameter	Test No., IHR-		
	189	190	184
Temperature, °C	850	850	850
Pressure, psig	1000	1000	1000
Coal	hvab	hvab	lignite
Coal rate, lb/hr	12.94	13.01	13.23
Feed gas rate, scfh	166.9	165.1	162.6
Hydrogen, vol %	52.4	49.4	50.1
Methane, vol %	44.0	46.0	47.8
Nitrogen, vol %	3.4	4.3	1.8
Total scf/lb	12.90	12.69	12.29
Hydrogen, scfh	87.5	81.6	81.5
Hydrogen, scf/lb	6.76	6.27	6.16
Product gas, scfh	191.4	184.2	204.1
Hydrogen, vol %	25.1	23.3	27.9
Methane, vol %	68.0	69.2	57.5
Ethane, vol %	0.2	0.2	0.1
Carbon monoxide, vol %	2.7	2.6	6.3
Carbon dioxide, vol %	0.8	0.6	5.9
Nitrogen, vol %	3.0	3.9	2.1

Table I. Continued

Parameter	Test No., IHR–		
	189	190	184
Product yield, Methane	4.38	3.98	3.25
Ethane, scf/lb	0.03	0.03	0.02
CO, scf/lb	0.37	0.34	1.03
CO_2, scf/lb	0.12	0.07	0.99
Feed H_2 reacted, scf/lb	3.05	2.99	2.01
Char residue, lb/lb	0.714	0.714	0.494
Condensed liquid, lb/lb			
Water	0.008	0.033	0.102
Oil	0.020	0.017	0.018
Residue moisture, lb/lb	0.009	0.018	0.020
Conversion, maf coal	33.1	30.6	50.7
Carbon, wt %	23.9	22.0	32.1
Hydrogen, wt %	66.4	60.9	77.4
Sulfur, wt %	47.5	38.6	51.0
Nitrogen, wt %	30.1	19.0	51.0
Recovery, overall	95.7	96.7	93.5
Carbon, wt %	98.8	96.9	97.6
Hydrogen, wt %	98.7	106.9	97.0
Ash, wt %	109.1	89.6	101.8

Parameter	Test No., IHR–					
	161	162	163	164	191	192
Temperature, °C	900	900	900	900	725	650
Pressure, psig	1000	1500	2000	1200	1000	1000
Coal	Ill. #6	Ill. #6	Ill. #6	Ill. #6	Ill. #6	Ill. #6
Coal rate, lb/hr	10.53	12.31	12.77	11.78	12.19	12.04
Feed gas rate, scfh	157.2	156.2	158.8	165.1	169.8	188.9
Hydrogen, vol %	54.5	48.8	52.0	50.1	56.5	60.1
Methane, vol %	44.5	49.3	46.1	47.7	42.8	39.4
Nitrogen, vol %	1.0	1.8	1.9	2.2	0.7	0.5
Total scf/lb	14.93	12.69	12.94	14.02	13.93	15.69
Hydrogen, scfh	85.6	76.2	82.6	82.7	96.0	113.5
Hydrogen, scf/lb	8.14	6.19	6.73	7.02	7.87	9.43
Product gas, scfh	182.3	185.3	190.1	199.4	190.4	202.0
Hydrogen, vol %	27.9	20.4	20.0	21.9	43.0	52.2
Methane, vol %	68.6	75.0	73.8	72.8	52.0	45.3
Ethane, vol %	trace	trace	0.2	trace	2.3	0.9
Carbon monoxide, vol %	1.4	1.9	2.2	2.4	1.6	0.9
Carbon dioxide, vol %	0.6	0.8	1.3	0.7	0.6	0.2
Nitrogen, vol %	1.4	1.7	2.3	1.9	0.4	0.4
Product yield, Methane	5.28	5.03	5.47	5.64	2.16	1.42
Ethane, scf/lb	trace	trace	0.03	trace	0.36	0.15
CO, scf/lb	0.24	0.29	0.34	0.41	0.25	0.15
CO_2, scf/lb	0.10	0.11	0.20	0.12	0.09	0.03
Feed H_2 reacted, scf/lb	3.31	3.12	3.63	3.31	1.15	0.67
Char residue, lb/lb	0.622	0.658	0.663	0.653	0.702	0.782

Table I. Continued

Parameter	Test No., IHR–					
	161	162	163	164	191	192
Condensed liquid, lb/lb						
Water	0.079	0.068	0.048	0.063	0.043	0.038
Oil	0.032	0.011	0.006	0.010	0.062	0.035
Residue moisture, lb/lb	0.012	0.013	0.013	0.014	0.010	0.009
Conversion, maf coal	38.0	37.6	36.6	35.5	31.4	24.3
Carbon, wt %	29.8	27.8	26.3	27.8	25.1	19.1
Hydrogen, wt %	70.1	70.3	72.0	70.6	55.5	47.0
Sulfur, wt %	43.8	50.9	63.8	51.2	51.8	42.1
Nitrogen, wt %	26.2	30.5	37.9	34.1	17.4	13.7
Recovery, overall	94.6	97.2	99.0	98.7	95.8	94.5
Carbon, wt %	96.5	96.1	100.0	99.5	97.1	95.1
Hydrogen, wt %	104.7	106.0	105.4	106.7	100.6	97.3
Ash, wt %	102.4	100.0	100.6	102.1	101.6	105.7

[a] Feed coal particle size range is 100 × 200 mesh.
[b] Reactor length, 3 ft.
[c] Feed coal particle size range is 100 × 200 mesh.

Table II. Typical Analyses of Coals[a] Used in This Study

	Pittsburgh Seam hvab Coal	Illinois No. 6 hvcb Coal	N. Dakota Lignite
Proximate analysis			
Moisture	1.2	1.4	7.8
Volatile matter	36.4	36.8	39.7
Fixed carbon	56.7	55.9	46.9
Ash	5.7	5.9	5.6
Ultimate analysis (dry basis)			
Carbon	79.09	75.45	64.64
Hydrogen	5.22	5.12	4.48
Nitrogen	1.60	1.72	0.76
Sulfur	1.10	1.32	0.76
Oxygen by difference	7.22	10.41	23.29
Ash	5.77	5.98	6.07
Total	100.00	100.00	100.00

[a] Hvab coal from U.S. Bureau of Mines experimental mine, Bruceton, Pa. Hvcb coal from Orient #3 mine, Freeman Coal Co., Waltonville, Ill. Lignite from Baukol-Noonan mine, Burke Co., N. Dakota.

Production of High Btu Gas. The feasibility of producing a gas having a heating value of 900 or more Btu/standard ft³ (scf) (after cleanup methanation) was established by several experiments designed to simulate the operation of an integrated hydrane reactor which consists of two stages (1). In such an integrated reactor, the hydrogen is first fed to a fluid bed where it reacts with char produced by the FDP reactor. The product gas from this fluid bed is the feed gas to the top of the FDP reactor, and it consists of about 50 vol % methane; the remainder is hydrogen plus a small amount of carbon monoxide. Thus the composition of the feed gas to the experimental isolated FDP reactor was adjusted to simulate the fluid-bed product gas from an integrated operation.

Table III compares carbon conversion, gas composition, and gas yields for specific experiments; the results are used to guide an economic evaluation of the Hydrane process (2). These results show that the goal of producing a high Btu gas can be achieved at 1000 psig and higher. For all three coals evaluated, Pittsburgh Seam hvab coal, Illinois No. 6 hvcb coal, and ignite, the gas produced after methanation to reduce CO to an acceptable level could be substituted for natural gas. Of course with lignite the higher oxygen content results in higher yields of CO, and this gas will therefore require more methanation than the product gas from the Illinois or Pittsburgh Seam bituminous coals. However even with lignite the fraction of the total methane that is produced directly rather than by methanation is greater than can be achieved by other processes using bituminous coal. In an actual plant where hydrogen is produced from the residual char, the catalytic water–gas shift reaction ($CO + H_2O \rightarrow H_2 + CO_2$) would not be carried to completion. Instead, as the base case analysis of the feed gas to the dilute phase indicates, some CO would be left in the hydrogen, resulting in a somewhat higher CO concentration in the raw product gas from the FDP reactor. This additional CO would result in increased hydrogen consumption during methanation and thereby lower the hydrogen content and increase the

Table III. Production of Pipeline-Quality Gas in FDP Reactor

Parameter	\multicolumn{5}{c}{Test No., IHR—}				
	Base	156	176	151	166
Pressure, psig	1000	1000	1000	1100	1200
Coal	hvab	hvab	hvab	hvab	hvab
Feed gas: coal ratio, scf/lb	11.1	12.3	12.5	11.9	12.3
Feed gas composition, vol %					
Hydrogen	46.1	49.0	48.0	48.0	49.2
Methane	47.5	49.4	49.4	49.2	48.7
Nitrogen	0.0	1.6	2.2	2.6	2.1
Carbon monoxide	4.0	0.0	0.0	0.0	0.0
Carbon dioxide	1.7	0.0	0.0	0.2	0.0
Carbon conversion, wt %	20.0	25.0	24.0	25.3	25.6
Product gas (water-free): coal ratio, scf/lb	14.8	13.4	13.9	14.1	13.3
Product gas comp. (water-free)					
Hydrogen	21.4	22.4	22.8	22.1	22.7
Methane	68.8	71.4	71.6	71.7	72.2
Ethane	0.0	0.5	0.2	0.1	0.1

Table III. Continued

Parameter	Test No., IHR–				
	Base	156	176	151	166
Product gas comp. (water free) (continued)					
Carbon monoxide	4.2	3.2	2.5	2.5	2.3
Carbon dioxide	1.3	0.7	0.7	0.5	0.5
Nitrogen	1.0	1.4	1.9	2.7	2.3
Hydrogen sulfide		0.4	0.2	0.4	0.2
Methane: hydrogen in product	3.21	3.19	3.14	3.24	3.18
Heating value, as-received, Btu/scf	779	817	812	802	815
Heating value with 4% CO methanation, Btu/scf	927	918	908	903	914
Percent methane equivalent ($CH_4 + C_2H_6$) made directly	94.2	94.7	94.7	94.7	94.8

Parameter	Test No., IHR—					
	160	165	154	157	158	172
Pressure, psig	1500	1500	2000	2000	2000	2000
Coal	hvab	hvab	hvab	hvab	hvab	hvab
Feed gas: coal ratio, scf/lb	13.1	12.3	12.5	12.4	12.8	13.2
Feed gas composition						
Hydrogen	53.8	51.3	52.5	49.9	51.8	50.6
Methane	43.4	47.0	46.4	48.4	46.6	46.4
Nitrogen	2.6	1.7	1.0	1.7	1.6	3.0
Carbon monoxide	0.0	0.0	0.0	0.0	0.0	0.0
Carbon dioxide	0.2	0.0	0.0	0.0	0.0	0.0
Carbon conversion, wt %	24.2	23.3	25.1	30.0	25.0	28.0
Product gas (water-free): coal ratio, scf/lb	13.5	13.4	14.0	13.9	14.0	13.9
Product gas comp. (water-free)						
Hydrogen	19.7	21.7	19.8	18.1	18.0	20.4
Methane	75.2	73.4	75.9	79.0	78.7	75.8
Ethane	0.3	0.1	0.1	0.1	0.1	0.1
Carbon monoxide	1.4	2.1	2.2	0.5	1.1	0.7
Carbon dioxide	0.8	0.6	0.4	0.4	0.4	0.2
Nitrogen	2.4	2.0	1.5	1.7	1.6	2.6
Hydrogen sulfide	0.1	0.2	0.1	0.2	0.2	0.2
Methane: hydrogen in product	3.82	3.38	3.83	4.36	4.37	3.72
Heating value (as-received) Btu/scf	835	823	842	863	862	839
Heating value with 4% CO methanation, Btu/scf	928	916	936	948	949	920
Percent methane equivalent ($CH_4 + C_2H_6$) made directly	95.0	94.8	95.0	95.2	95.2	95.0

Table III. Continued

Parameter	Test No., IHR—			
	164	162	163	184
Pressure, psig	1200	1500	2000	1000
Coal	hvcb	hvcb	hvcb	lignite
Feed gas: coal ratio, scf/lb	14.0	12.7	12.9	12.3
Feed gas composition				
Hydrogen	50.1	48.8	52.0	50.1
Methane	47.7	49.3	46.1	47.8
Nitrogen	2.2	1.8	1.9	1.8
Carbon monoxide	0.0	0.0	0.0	0.2
Carbon dioxide	0.0	0.1	0.0	0.0
Carbon conversion, wt %	27.8	27.8	26.3	32.1
Product gas (water-free): coal ratio, scf/lb	16.9	15.0	15.5	10.6
Product gas comp. (water-free)				
Hydrogen	21.9	20.4	20.0	27.9
Methane	72.8	75.0	73.8	57.5
Ethane	trace	trace	0.2	0.1
Carbon monoxide	2.4	1.9	2.2	6.3
Carbon dioxide	0.7	0.8	1.3	5.9
Nitrogen	1.9	1.7	2.3	2.1
Hydrogen sulfide	0.3	0.2	0.2	0.1
Methane: hydrogen in product	3.32	3.68	3.69	2.06
Heating value (as-received) Btu/scf	818	833	824	695
Heating value with 4% CO methanation, Btu/scf	914	928	925	902
Percent methane equivalent ($CH_4+C_2H_6$) made directly	94.8	94.9	94.9	90.1

heating value of the final product gas. The heating value of the final product gas was therefore calculated based on a constant 4 vol % CO in the raw, dry product gas from the FDP reactor.

Examination of these FDP results indicates that the following controllable parameters determine whether the raw product gas will, after methanation of the 4 vol % CO, have a heating value of at least 900 Btu/scf: a) per cent methane in the feed gas to the FDP reactor, b) gas-to-coal feed ratio, c) reactor pressure, d) coal residence time (reactor length).

Fortunately the combination of above variables required to produce 900+ Btu gas is easy to achieve in practical reactor systems. For example, the reactor can operate at gas transmission-line pressures, both the gas:coal feed ratio and the methane concentration in the feed gas allow operation at carbon conversion levels resulting in balanced plant operation, and a sufficient coal residence time was achieved in an FDP reactor only 5 ft long. (Balanced operation means the overall plant produces no surplus char. To achieve balanced operation, the fraction of carbon in the coal converted to methane is regulated so that the remaining carbon is just sufficient to produce process hydrogen and plant power.) Increases in the per cent methane in the feed gas to the FDP reactor, at constant CO concentration, will further increase the heating value of the final product gas.

In our experiments the reactor pressure shell was pressurized with the feed gas. Because this gas was in direct contact with the reactor electrical heating elements, carbon deposition from methane cracking and subsequent shorting of the electrical elements became a problem when the methane concentration in the feed gas was higher than 50 vol %. This artificial limitation would not exist in a commercial reactor system where no electric heating elements would be used. Methane concentrations above 50 vol % would be generated by the fluid-bed stage of the Hydrane reactor system, and therefore higher methane concentrations would be possible to the FDP reactor. As examples, Pyrcioch and Linden (5) studied the fluid-bed hydrogasification of a char produced by low temperature pretreatment and achieved methane concentration over 50 vol %; Lewis and co-workers (9) reported methane concentrations over 60 vol % from direct moving-bed hydrogasification of chars produced by hydrogasification in an FDP reactor. Thus the results presented here must be regarded as conservative, and higher methane concentration product gases could be produced in commercial FDP reactors where the methane concentration in the intermediate feed gas to the FDP reactor is not limited by artificial constraints.

Product Yields and Distribution. The major products from the dilute-phase hydrogasification of raw coal are gas and char plus smaller amounts of organic liquid products and water. The liquid yield measurement is inaccurate because of the relatively small amount of liquids formed and the difficulty of their quantitative recovery. For all the experiments reported in Table I, measured yields of organic liquids varied from less than 0.01 to 0.06 lb/lb coal. Attempts to correlate the organic liquid yields with reactor parameters thought to have the greatest effect on these yields such as reactor wall temperature, hydrogen partial pressure, and gas-phase residence time have been unsuccessful. In a baseload, pipeline gas plant the organic liquid production will be substantial even at the lowest yields measured, so additional work is now going on to characterize these organic liquids.

Water is produced by both vaporization of moisture in the coal and reaction of hydrogen with oxygen in the coal. Recoveries of water both from condensers and as moisture on the char ranged from 0.01 to 0.08

lb/lb of coal fed. However as Table I indicates, the water recoveries measured for the Illinois No. 6 hvcb coal and the single run made with lignite were higher because of the higher oxygen contents of these feeds. For the Illinois coal, water yields varied from about 0.05 to 0.09 lb/lb coal. These water-yield data indicate that much of the oxygen either is present as bound water or combines with hydrogen to form water during hydrogasification.

Char Yields and Desulfurization. In the overall Hydrane process, the char from the FDP reactor will be further converted in a fluid-bed reactor which is in series with the FDP reactor. The yield of char from the FDP reactor depends on the carbon conversion level as shown in Figure 1.

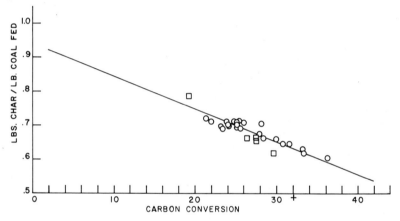

Figure 1. Dependence of FDP reactor char yield on carbon conversion. + is Lignite, ○ is Pittsburgh Seam hvab, and □ is Illinois No. 6 hvcb.

Sulfur is eliminated from the char during hydrogasification as H_2S and the degree of elimination is related to the carbon conversion as shown in Figure 2. The scatter may be a result of the influence of parameters other than carbon conversion and/or the inaccuracies in the sulfur determinations. The important point to be demonstrated is that the coal sulfur is extremely reactive under hydrogasification conditions as seen by the coal residence time in the FDP reactor of about 1-2 sec. In fact, Figure 2 indicates that the sulfur in the coal is more reactive than the carbon in the coal. The Pittsburgh Seam hvab coal contains approximately equal amounts of pyritic and organic sulfur. However, the char has not been tested to indicate whether either type is selectively removed during free-fall hydrogasification. In the integrated Hydrane process the char spends additional residence time in a fluid bed at higher hydrogen partial pressures than exist in the FDP reactor, so additional char desulfurization will occur in the fluid bed. In preliminary experiments with an integrated

FDP fluid-bed reactor system, the sulfur removal from the Pittsburgh Seam coal has been about 85%. These results are encouraging because they indicate that char from the Hydrane reactor may be an acceptable fuel to provide the plant's energy and steam requirements without complicated sulfur removal systems and without exceeding air quality restrictions on atmospheric release of sulfur compounds.

Scaleup of the FDP Reactor. The FDP reactor has two important functions: it must convert the coal to a nonagglomerating char for the subsequent fluid bed and it must convert enough carbon to methane so that the FDP product gas is, after acid gas removal and light methanation, an acceptable pipeline gas.

In the 3-inch id FDP reactor used in our experiments, the coal particles are heated to reaction temperature in the reactor by mixing with the preheated feed gas and by heat transfer from the hot walls of the FDP reactor. However heat transfer analysis of larger reactors (*11*) indicates that as the reactor diameter is increased the amount of heat transfer from hot reactor walls to the particles inside becomes negligible. Therefore in larger diameter reactors, the coal particles can be raised to reaction temperature only by mixing with the hot methane–hydrogen mixture shunted from the fluid bed. Calculations indicate that the mixing temperature of the hot gas and coal at the top of a large FDP reactor will be about 480°–540°C. It is therefore important to evaluate the

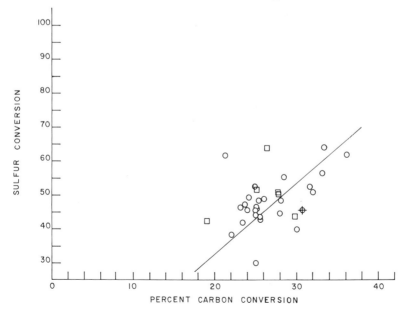

Figure 2. Sulfur elimination in FDP reactor. + is Lignite, ○ is Pittsburgh Seam hvab, and □ is Illinois No. 6 hvcb.

FDP hydrogasification behavior at these relatively low temperatures. This is difficult to do in the present 3-inch id FDP reactor because the coal quickly heats to the reactor wall temperature and, if the wall temperature is below 725°–800°C, the coal adheres to the reactor walls and eventually plugs the reactor. However, the coal that did not contact the walls passed through the reactor and was collected, and its conversion and caking properties were determined. Results of these lower reactor wall-temperature experiments are shown in Figures 3 and 4. The effects of temperature on both the volatile matter and the carbon conversion of the FDP reactor char are shown at the reduced wall temperatures.

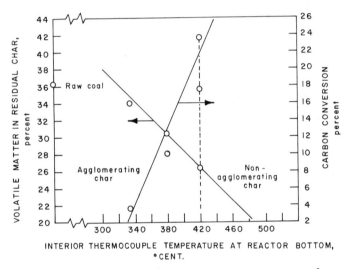

Figure 3. *Effect of estimated particle temperature on carbon conversion and volatile matter–reactor bottom temperature*

Figure 3 shows an interior thermocouple temperature at the reactor bottom and Figure 4 shows the average reactor wall temperature because the actual particle temperature is not known. The true average particle temperature is probably between the interior thermocouple temperature and the reactor wall temperature. Also shown in Figures 3 and 4 is the temperature boundary above which the char is not agglomerating when tested in a fluid bed with hydrogen at 1000 psig and 900°C. Chars produced at reactor wall temperatures below the boundary temperatures agglomerated when tested at the above conditions. Thus if one conservatively assumes that the particle temperature is close to the wall temperature, it appears that mixing the hot methane–hydrogen mixture produced in the fluid-bed reactor with the coal at the top of the FDP reactor will produce an acceptable nonagglomerating char if the char temperature reaches 580°C, even for a residence time of only 1-2 sec.

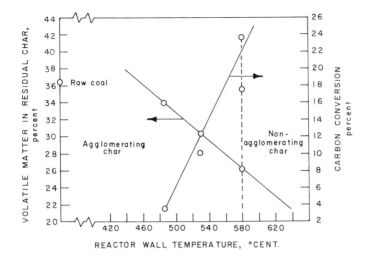

Figure 4. Effect of estimated particle temperature on carbon conversion and volatile matter–reactor wall temperature

Previous reaction rate analyses of FDP reactor data (*10*) at higher hydrogen partial pressures (1500 to 3000 psig) and at reactor wall temperatures of 725° and 900°C indicated that the conversion of carbon in raw coal occurs in three stages, and each stage has greatly different reactivities toward hydrogen. At the short (1-2 sec) particle residence times in the FDP reactor at 3000 psig and with wall temperatures of 900°C, all of the first-stage (the most reactive) carbon behaved as if it were converted instantaneously. However the conversion of the second-stage carbon varied with reactor conditions and this variation was correlated by the rate equation:

$$U_T \frac{dx}{dL} = k p_{H_2} (1-x) \tag{1}$$

where U_T is the average terminal velocity of the particles in ft/hr, k is the rate constant per atm-hr, p_{H_2} is the partial pressure of hydrogen in atm, L is the location in the reactor (the distance from the coal inlet in ft), and x is the fractional carbon conversion level. The fraction of carbon that behaved as though it were instantaneously converted was denoted by E and was determined by finding the value that allowed the best fit of experimental data with the integrated form of Equation 1:

$$\int_E^x \frac{dx}{p_{H_2}(1-x)} = \frac{k}{U_T} L \tag{2}$$

At conditions where the conversion of second-stage carbon, x-E, was small, errors or variations in either x or E caused large fluctuations in the value of k, making a kinetic analysis of the data difficult. This occurs where the total carbon conversion is mostly from the instantaneous carbon reaction because conditions are not severe enough to activate a substantial fraction of the second-stage carbon. As an example of severe conditions, operation at 3000 psig, 900°C wall temperatures, and hydrogen feed gas allowed total carbon conversions ranging from 0.40 to 0.50 in the FDP reactor. At these conditions the value of E is approximately 0.14 and the spread between x and E is sufficiently large to allow a reasonable determination of k. For the experiments reported in reference 10, E varied from about 0.15 to 0.20.

At the conditions reported in this paper where the total pressure is closer to 1000 psig and the feed gas to the FDP reactor is an approximately equimolar mixture of hydrogen and methane, the total carbon conversions are closer to the fraction of carbon that instantaneously reacts and kinetic interpretation is even more difficult. Therefore the kinetic analysis is not yet complete. However for the purposes of FDP reactor simulation, a mathematical model was used that assumed all the carbon reacts at a rate dictated by Equation 1 rather than assuming a portion of this carbon reacts instantaneously. This assumption is felt to be conservative because it does not allow for the fraction of carbon that may react at a considerably faster rate than the final amount of carbon conversion which was used to evaluate the rate constant k. The temperature dependency of k used for our initial reactor simulation studies *(11)* has been reported *(1)*. While the more detailed kinetic analysis may result in a modified rate equation, the results of our simulation study *(11)* indicate that radiant heat transfer plays a dominant role in small FDP reactors such as the one used in this study. Because the effect of radiant heat transfer from the reactor walls diminishes as the diameter of the reactor increases, temperature profiles in commercial reactors will be considerably different from those existing in our present 3-inch id FDP reactor; this indicates the necessity of using larger diameter pilot plants to obtain reliable scaleup data.

Literature Cited

1. Feldmann, H. F., Wen, C. Y., Simons, W. H., Yavorsky, P. M., AIChE, Nat. Meetg., *71st*, Feb. 20-23, 1972.
2. Wen, C. Y., Li, C. T., Tscheng, S. H., O'Brien, W. S., AIChE, Nat. Meetg., *65th*, Nov. 26-30, 1972.
3. Dent, F. J., Blackburn, W. H., Millet, H. C., Trans. Inst. Gas Eng. (1938-1939) **88**, 150.
4. Channabasappa, K. C., Linden, H. R., Ind. Eng. Chem. (1956) **48**, 900.
5. Pyrcioch, E. J., Linden, H. R., Ind. Eng. Chem. (1960) **52**, 590.

6. Lee, B. S., Pyrcioch, E. J., Schora, F. C., Jr., ADVAN. CHEM. SER. (1967) **69**, 104.
7. Wen, C. Y., Huebler, J., *Ind. Eng. Chem.* (1965) **4**, 142, 147.
8. Hiteshue, R. W., *Proc. Synthet. Pipeline Gas Symp.*, Pittsburgh, Nov. 15, 1966, p. 13.
9. Lewis, P. S., Friedman, S., Hiteshue, R. W., ADVAN. CHEM. SER. (1967) **69**, 50.
10. Feldmann, H. F., Simons, W. H., Mima, J. A., Hiteshue, R. W., *Amer. Chem. Soc., Div. Fuel Chem., Prepr.* **14** (4), 1 (Chicago, September 1970).
11. Feldmann, H. F., Simons, W. H., Wen, C. Y., Yavorsky, P. M., *Intern. Congr. Chem. Eng., 4th*, Sept. 11-15, 1972.

RECEIVED May 25, 1973.

9

Chemistry and Physics of Entrained Coal Gasification

R. L. ZAHRADNIK

National Science Foundation, Washington, D. C. 32701

R. J. GRACE

Bituminous Coal Research, Inc., 350 Hochberg Rd., Monroeville, Pa. 15146

Pulverized coal, when entrained in a stream of hot, high pressure hydrogen-containing gas, can be converted in good yields to methane and other combustible gases. Methane yields can be related to process variables such as temperature, hydrogen partial pressure, and coal rank. The 100 lbs/hr internally fired gasifier was designed so that the momentum of the coal-feed stream injected into the hot synthesis gas created recirculating flows that reintroduced a portion of the methane into the higher temperature region of the gasifier. There, steam re-forming of the methane occurred with resultant decrease in overall methane production. Design objectives for entrained gasifiers are proposed which should minimize the methane losses through reforming in secondary flows.

Key features of the Bi-Gas process for producing synthesis gas (1, 2) are illustrated in Figure 1. Fresh coal is introduced into the upper section (stage 2) of a two-stage gasifier at system pressures of 70-100 atm. Here it contacts a rising stream of hot synthesis gas produced in the lower section (stage 1) of the gasifier and is partially converted to methane and more synthesis gas. The residual char is swept out of the gasifier together with the gas; the char is separated from the gas stream and returned to the bottom section of the gasifier. Here the char is completely gasified under slagging conditions by reaction with oxygen and steam, producing both the synthesis gas required in the upper section of the gasifier and the heat needed to complete the endothermic reactions.

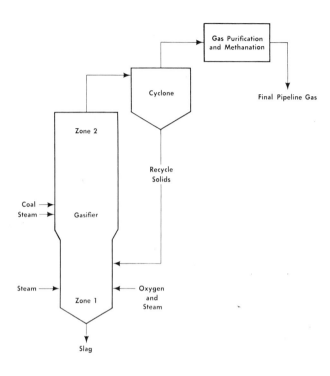

Figure 1. Simplified flow diagram for two-stage superpressure gasifier

To meet pipeline specifications the product gas requires further processing. It is cleaned and subjected to partial water–gas shift to adjust the H_2/CO ratio; it is scrubbed to remove acid gases (CO_2, H_2S); finally it is subjected to catalytic methanation to increase the heating value to pipeline quality.

The basic component of the process is the entrained two-stage gasifier. The major emphasis of the BCR program to date has been on developing data sufficient to optimize stage 2 of the gasifier. Initial experiments were carried out with coal slurries in rocking autoclaves at 3000-4000 psig and 1380°-1400°F (3). These experiments showed that large amounts of methane could be produced from the contact of coal with superheated steam. However the batch-size tests involved relatively slow heating rates and long residence times. Consequently the results could not be applied

Table I. Range of Operating Conditions for 100 lbs/hr PEDU Tests

Coal feed rates, lbs/hr	43–108
System pressure, psig (atm)	220–1420
	(15–96.5)
Outlet temperature, °F (°C)	1375–2160
	(746–1180)
Hydrogen partial pressure, psig (atm)	56–308
	(3.8–21)
Total steam/coal ratios	0.90:1–2.78:1
Total carbon gasified, %	32–68
Total Btu in gas:Btu in coal, %	38–85
Residence times, sec	3–22

directly to an integrated entrained gasifier, and we sought data from experiments under more realistic conditions.

Subsequent tests were continuous and involved a short coal–steam–synthesis gas contact time with rapid heating. Over 100 experiments were conducted under conditions simulating those of stage 2, using a 5 lbs/hr continuous-flow reactor (CFR). Lignite, a Wyoming subbituminous coal and Pittsburgh seam high volatile bituminous coal were tested. These experiments showed conclusively that appreciable amounts of methane could be produced during short contact times of 2-20 sec between steam, synthesis gas, and fresh coal at about 1000 psi and 1750°F (4).

The experiments carried out in the 5 lbs/hr unit involve the simultaneous heating of the simulated stage 1 gas, the superheated steam, and the fresh coal. Because of the limitations of the equipment, the reaction conditions did not exactly duplicate those expected in the integrated gasifier. Nevertheless the results warranted the construction of a process and equipment development unit (PEDU) in which fresh coal and steam could be contacted with hot stage 1 gas under conditions that more closely duplicate those in stage 2. The design features of this PEDU are given elsewhere (1, 5, 6, 7). Nearly 60 individual experiments were conducted using the same series of coals as were used in the CFR. The range of operating conditions for these tests is reported in Table I; the range of results is given in Table II.

Initial results from the PEDU for lignite were reported earlier (2, 8). This paper presents data for the gasification of Pittsburgh seam coal in the PEDU and discusses these data on the basis of the chemistry and physics of entrained gasification.

Gasification Chemistry

The physical and chemical processes which take place between the hot synthesis gas from stage 1 and the fresh coal and steam in stage 2

Table II. Range of Results for 100 lbs/hr PEDU Tests

Coal	Coal Feed Rate, lbs/hr	Methane Yield, %	Total Carbon Gasified, %	Preformed Methane, %
Design Basis	100	15.0	33	62
Lignite	62–108	12–20	32–68	30–79
Elkol	43–104	18–23	39–64	47–86
Pittsburgh	50–77	16–26	33–57	61–84
Lower Kittanning	64–66	14–17	34–38	59–68

are complex, and any attempt to model them must be regarded as approximate. Nonetheless it is possible to develop reasonable correlations in terms of certain basic gasifier variables suggested by the gasification chemistry.

As a result of the very rapid heating of the coal, a significant devolatilization takes place in milliseconds (9). This devolatilization step produces various gases including hydrogen and methane. The remainder of the gasification process may be characterized by the carbon–hydrogen reaction,

$$C + 2H_2 \rightarrow CH_4 \tag{1}$$

and by the carbon–steam reaction:

$$C + H_2O \rightarrow CO + H_2 \tag{2}$$

The overall methanation has been described by Moseley and Paterson (10, 11, 12) as consisting of three steps. The first step is the rapid devolatilization of coal which produces, in addition to volatile products, an active carbon species. This active carbon reacts in the second step either with hydrogen to form more methane or with itself in a crosslinking polymerization to form an inactive char. The third step involves the slow reaction of hydrogen with the inactive char.

In stage 2 of the BCR two-stage process this third step is relatively unimportant. Zahradnik and Glenn (13) have shown that the direct methanation of coal in stage 2 can be described adequately as a two-step process which is independent of residence times greater than a few seconds. On this basis it is possible to relate the yield of methane, MY, expressed as the fraction of the carbon in coal appearing as methane, to hydrogen partial pressure in the following way

$$MY = \frac{a + b(P_{H_2})}{1 + b(P_{H_2})} \tag{3}$$

where a and b are kinetic parameters.

Zahradnik and Glenn (13) were able to correlate the CFR data, the data of Moseley and Paterson, and the data obtained by the U.S. Bureau of Mines with this expression. Data obtained in the 100 lbs/hr PEDU for lignite were also shown to be correlated by Equation 3 (8), demonstrating its validity for larger sized equipment.

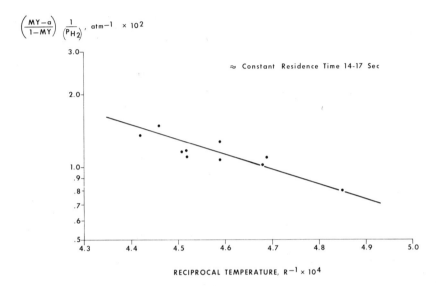

Figure 2. Methane yield per atmosphere of hydrogen partial pressure as a function of reciprocal outlet temperature at constant residence time for Pittsburgh seam coal

For data-correlation, Equation 3 can be written as:

$$\left(\frac{MY - a}{1 - MY}\right) \frac{1}{(P_{H_2})} = b \tag{4}$$

On the basis of earlier tests, a is taken to be 0.08 and is assumed to depend on temperature in an Arrhenius manner. Thus a plot of the natural logarithm of the left side of Equation 4 as a function of reciprocal temperature should yield a straight line. Figure 2 is such a plot for Pittsburgh seam coal under comparable physical and geometrical configurations and for residence times between 14 and 17 sec. The correlation is quite good. A complete tabulation of the data on which Figure 2 is based is given elsewhere (1). The experimental and analytical techniques used to obtain these data were reported in a previous publication by Grace et al. (8). Not all the data points from the PEDU experiments with Pittsburgh seam coal fell on the correlating line of Figure 2, however. A more detailed explanation of their behavior is in order.

Methane Decomposition

Experiments in the externally heated 5 lbs/hr CFR showed that methane, once formed, did not decompose under the stage 2 simulation achieved with this unit. However, because of the higher mixing temperatures attained in the PEDU, such decomposition is possible. To test whether methane does decompose, a stream of methane was injected into the simulated stage 1 gas where it experienced partial decomposition. The exact nature of the methane destruction is not clear. However it is most likely that the steam in the stage 1 gas promotes the reforming reaction:

$$CH_4 + H_2O \rightarrow CO + 3H_2$$

Analysis of the material balance data from the methane decomposition tests suggests that the latter reaction is occurring. One test was carried out at 200 psig to permit observation of the mixing temperature by an ultraviolet pyrometer. The results can be explained by assuming that methane decomposes at a rate proportional to its concentration, *i.e.*,

$$\frac{d[CH_4]}{dt} = k[CH_4]$$

Although it is likely that the reaction rate is influenced by steam, hydrogen, and carbon monoxide partial pressures, these did not vary significantly during the tests, and their effect cannot be determined at this time. Integration of the rate expression gives:

$$\ln[1 - f] = -k\tau \qquad (5)$$

where

τ = residence time
f = fraction of methane decomposed
k = reaction rate constant

The residence time of the individual tests was constant. Because the reaction rate constant is temperature dependent, an Arrhenius plot of $\ln(\ln[1 - f])$ *vs.* the reciprocal of the observed mixing-zone temperature should yield a straight line. This is indeed the case, as shown in Figure 3. The temperature effect on the rate of methane decomposition is quite pronounced, corresponding to an activation energy of 30 kcal/gram-mole.

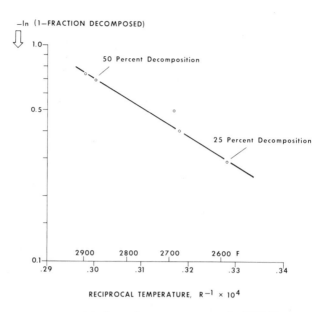

Figure 3. Methane decomposition in the PEDU

The fact that methane injected into the PEDU decomposes to a certain extent suggests that methane formed directly from coal could also decompose. Thus the methane yield predicted by Equation 3 would have to be modified as in Equation 5. This gives the following equation for methane yield:

$$MY = \left(\frac{a + b(P_{H_2})}{1 + b(P_{H_2})}\right) e^{-k\tau} \qquad (6)$$

This equation indicates that methane yield depends on residence time but in an unusual and unexpected way.

Because both b and k depend on temperature, it is difficult to express the relationship of Equation 6 in a form convenient for graphical display. However certain first-order simplifications and approximations can be made. Thus if we make the approximation, $e^{-k\tau} \cong 1 - k\tau$, Equation 6 can be written in the following form:

$$\left(\frac{MY - a}{1 - MY}\right)\left(\frac{1}{(P_{H_2})}\right) = b - \frac{k\tau[a + b(P_{H_2})]}{(P_{H_2})(1 - MY)} \qquad (7)$$

Further, if we note that the group $[a + b\ (P_{H_2})/(P_{H_2})]$ is relatively insensitive to the partial pressure of hydrogen in the range of experiments conducted, Equation 7 becomes

$$\left(\frac{MY - a}{1 - MY}\right)\left(\frac{1}{(P_{H_2})}\right) = b - k_1\left(\frac{\tau}{1 - MY}\right) \quad (8)$$

Thus a plot of

$$\left(\frac{MY - a}{1 - MY}\right)\left(\frac{1}{(P_{H_2})}\right) vs. \frac{\tau}{1 - MY}$$

should give a straight line provided temperature is constant. Figure 4 is such a plot for those Pittsburgh seam coal tests with an exit temperature of 1720°–1800°F. The trend is unmistakable. Although Equation 8 is the result of a number of assumptions and mathematical simplifications, it does provide a format for displaying and correlating the PEDU methane-yield data. In addition these data do suggest that methane formed in the direct methanation process is destroyed in its passage through the remainder of the PEDU. The decomposition indicated by Equation 8 ranges from 10% at the low residence time tests to 25% at the high residence time tests.

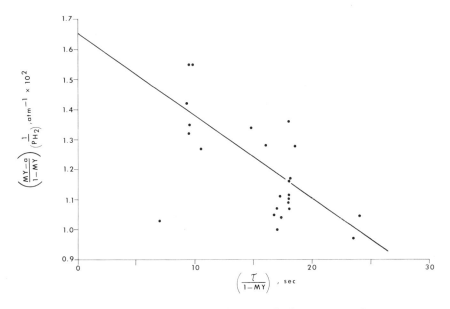

Figure 4. Methane yield per atmosphere of hydrogen partial pressure vs. residence time for Pittsburgh seam coal

If the correlating line in Figure 4 is extrapolated to zero residence time, a value is obtained for the b parameter in the methane yield equation: $b = 0.0165$. An extension to higher temperatures of the Arrhenius plot for b obtained by Zahradnik and Glenn shows this value to correspond to a temperature of 2240°F (*13*). This is probably a reasonable estimate of the mean reaction temperature of the methanation process taking place in the PEDU when the exit temperature is between 1720° and 1800°F.

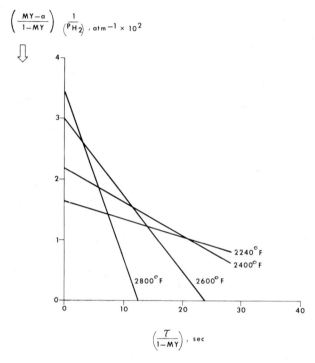

Figure 5. *Residence time and temperature effect on methane yield*

At 2240°F, the methane decomposition correlation (Figure 3) gives a value for $k\tau$ of 0.085. Since the residence time in the methane decomposition tests was 7.5 sec, a k value can be calculated.

$$k = \frac{0.085}{7.5} \text{ sec}^{-1} = 0.0113 \text{ sec}^{-1}$$

A k value can also be obtained from the gasification tests because Equation 8 gives the slope of the correlating line of Figure 4 as

$$\text{slope} = k \left(\frac{a + b(P_{H_2})}{(P_{H_2})} \right)$$

The value of this slope is 0.000275. If $b = 0.0165$ and the average hydrogen partial pressure is 15 atm, the value of k can be calculated:

$$k = \frac{0.000275 \text{ sec}^{-1}}{0.08 + 0.0165 \ (15)} = 0.0126 \text{ sec}^{-1}$$
(15)

This value, 0.0126 sec⁻¹, is remarkably close to the value of 0.0113 sec⁻¹ at 2240°F estimated from the methane decomposition tests.

A complete and consistent model for methane production in the PEDU may now be given. The following equation can be written for any temperature and contact time:

$$\left(\frac{MY - a}{1 - MY}\right)\left(\frac{1}{(P_{H_2})}\right) = b - k\left(b + \frac{a}{(P_{H_2})}\right)\left(\frac{\tau}{1 - MY}\right) \quad (8)$$

For a given reaction temperature, b can be estimated from data given by Zahradnik and Glenn (13) and k estimated from Figure 3 and Equation 5. Then a plot of reduced methane yield per atmosphere of hydrogen partial pressure as a function of reduced residence time can be constructed, as in Figure 5. As temperature increases, the decomposition

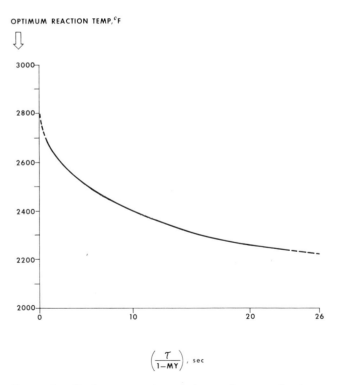

Figure 6. Optimum temperature for methane production vs. residence time

rate increases, thereby reducing the contact time required to destroy the methane. Moreover, for a given residence time, there is a best temperature at which to operate—best in the sense of yielding the highest methane yield per atmosphere of hydrogen partial pressure. This best temperature is shown as a function of residence time in Figure 6.

The results in Figures 5 and 6 are specific to the PEDU operation and involve numerous assumptions. Moreover, the temperature values are estimated reaction temperatures which combine the total effect of temperature profiles and gas mixing patterns. As in all the correlations presented here, they should be regarded as phenomenological and suggestive—not as the consequences of actual mechanisms. Nonetheless, they do underscore the effect of residence time and temperature on the yield from stage 2.

Steam–Carbon Reaction

The yield of carbon oxides from coal in stage 2 has been attributed to the steam–carbon reaction and water–gas shift.

$$C + H_2O \rightarrow CO + H_2$$
$$CO + H_2O \rightarrow CO_2 + H_2$$

The rate of production, under stage 2 conditions, is to a first-order approximation

$$\text{rate} = k' \frac{(P_{H_2O})}{(P_{H_2})}$$

where k' is an effective rate constant. Proper integration of this equation would have to take into account the temperature and composition paths which are unknown. However if outlet conditions are used to approximate the appropriate integrated equation, the following expression is obtained:

$$\frac{CY}{\left(\frac{(P_{H_2O})}{(P_{H_2})}\right)_\tau} = k' \qquad (9)$$

where CY = fraction of carbon in coal gasified to carbon oxides.

Figure 7 is a plot of the natural logarithm of the carbon oxide yield expression from Equation 9 *vs.* reciprocal outlet temperatures for all the data for Pittsburgh seam coal. The overall fit is fairly good, indicating that the various assumptions required to arrive at Equation 9 are not unreasonable.

Gasification Physics

The physical processes taking place during entrained gasification are as complex as the chemical ones. Complete modeling of the physics would have to include the expansion of the jet of coal into the hot synthesis gas, particle-particle collisions, particle heat-up, etc. Many of these physical processes have been examined by BCR in an attempt to understand entrained gasification. However, in light of the recirculation reformers set up by the coal feed, the most significant physical process affecting methane yield is the mixing between the coal feed and hot stage 1 gases.

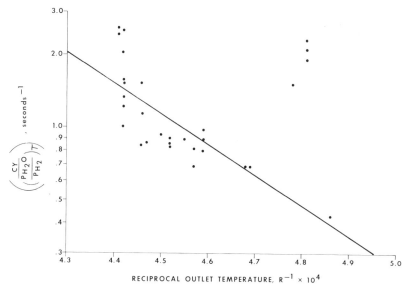

Figure 7. Carbon oxides yield group vs. reciprocal outlet temperature for Pittsburgh seam coal

Although it is not possible to predict or even to infer the exact mixing patterns in stage 2 of the PEDU, some insight into these flow patterns can be obtained by considering certain idealized conditions. If only the expansion of the coal feed jet into stage 2 is considered, the fluid mechanics of turbulent jets predict that the coal stream would strike the wall at a distance X_p (14, 15), where

$$X_p = 5.85\ L$$

and $2L$ is the stage 2 diameter (ft). In the absence of reaction or other influence from the stage 1 gases, the coal feed jet would strike the wall of the 8-inch diameter PEDU about 2 ft below its entrance.

As the coal-feed jet decreases from its nozzle velocity, it entrains surrounding fluid to conserve its axial momentum and thereby sets up recirculation currents. The mass rate of material recirculated, m_r, per mass rate of material fed, m_o, can be estimated from the following equation proposed by Thring (15):

$$\frac{m_r}{m_o} = \frac{0.47}{\theta} - 0.5 \qquad (10)$$

where

$$\theta = \frac{m_o}{L(\pi \rho_a G)^{1/2}}$$

ρ_a = density of surrounding fluid (lb/ft^3)
$G = m_o \times v_o$ = mass velocity at nozzle (ft-lb/sec^2)
Typical values for PEDU operation are:
  ρ_a = 0.6 lb/ft^3 (22 molecular weight gas, 1020 psi, 2700°R)
  m_o = 100 lb/hr = 1/36 lb/sec
  v_o = 50 ft/sec
  G = 50/36 ft-lb/sec^2

Hence

$$\theta = \frac{\frac{1}{36}}{\left[\frac{1}{3} \pi (0.6) \left(\frac{50}{36}\right)\right]^{1/2}} = \frac{1}{14.5}$$

And a typical recirculation ratio is:

$$\frac{m_r}{m_o} = \frac{0.470}{0.069} - 0.5 = 6.4$$

Operation at this ratio would cause the first 2 ft of stage 2 to be fairly well mixed. According to Thring, recirculation into the jet begins at a distance X_n (15),

$$X_n = 6.25 \, \theta \, L$$

which in this case is 1.7 in. Entrainment into the jet continues until a distance X_l,

$$X_l = 3.12 \, (0.94 + \theta) \, L$$

which is about 13 in. From this point onward disentrainment occurs, reaching a maximum at about 16 in. These dimensions are summarized on Figure 8 which is a schematic of the PEDU.

The high recirculation rate and the fact that entrainment is taking place in the region where the hot stage 1 gas enters stage 2 indicate that product decomposition could occur in the PEDU. As shown in the previous section, this does indeed occur.

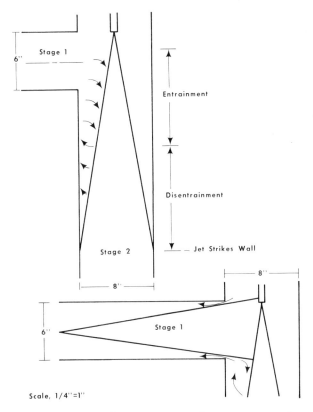

Figure 8. Typical entrainment, disentrainment distances in the PEDU

When stage 1 and stage 2 are operating in concert, the mixing patterns are unquestionably more complicated. The stage 1 zone is 18 in. long, and its diameter is 6 in. For these dimensions it might be expected, on the basis of cold jet mixing, that the stage 1 gases would strike the walls of the sidearm just prior to their entrance into stage 2. However it has been reported that the jet half-angle in a furnace flame is about 4 1/4°. At this angle the distance for the jet to strike the stage 1 chamber walls X_p, would be (9, 15):

$$X_p = \cot(8.5°) \, (3 \text{ inches}) = 20 \text{ inches}$$

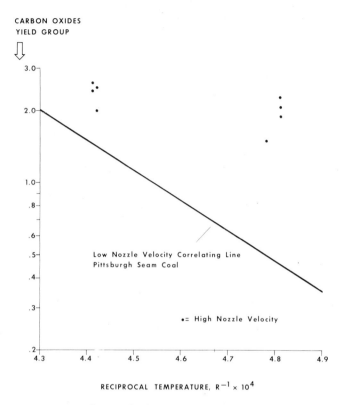

Figure 9. *Effect of high nozzle velocity on carbon oxides yield group for Pittsburgh seam coal*

Since this distance exceeds the length of the stage 1 zone, recirculation into stage 1 from stage 2 would be expected. From the previous arguments it must be concluded that this would include both product gas and char. In fact, considerable insufflation into stage 1 did occur because extensive slag deposits were observed along the entire bottom of this zone. With this concept of PEDU circulation in mind some tests were conducted in which hydrogen was used for coal transport. Low pressure operation was also used. In these cases nozzle velocities exceeded 100 ft/sec and in some cases approached 300 ft/sec. Because θ is proportional to $(m_o/v_o)^{1/2}$, and if all other variables are held constant, an increase in nozzle velocity to 200 ft/sec would decrease θ from the previously considered value by a factor of $1/2$ to $\theta = 0.035$. Using this value, the mass recirculated ratio then would become

$$\frac{m_r}{m_o} = 12.9$$

The locations of the entrainment and disentrainment areas would remain relatively the same, and ideally the jet would strike the stage 2 chamber walls at about the same 2-ft level.

Under such operation, one would expect a considerably higher degree of backmixing or recirculation although the average residence time of the gases in this part of the reactor would be dictated by overall flow rates. It is possible that the considerable backmixing leads to a higher inventory of char, particularly in the high temperature region exposed directly to stage 1 input. This would lead to proportionately higher char gasification (to carbon oxides) than in less well mixed tests.

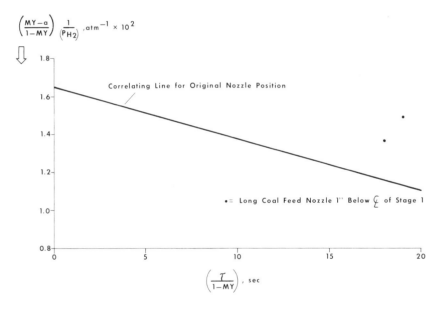

Figure 10. *Effect of location of coal feed nozzle on methane yield for Pittsburgh seam coal*

Figure 9 presents the results of the high nozzle velocity tests for Pittsburgh seam coal. The carbon oxide yields are indeed higher than expected. The methane yields for these tests correlated in the same way as did those for the lower nozzle-velocity tests and are included in Figure 4. This is expected because the reactivity of char to direct methanation by hydrogen is considerably less than the active form of carbon produced in the initial heat-up of coal. Consequently, a higher char inventory, even in the high temperature region, would not produce higher methane yields.

Discussion

The results and correlations expressed above are specific to the geometry and conditions of the PEDU. They do not represent fundamental limitations on the entrained gasification of coal but are guidelines to a basic understanding of the process. Thus, if the flow patterns set up by the PEDU geometry result in destruction of a portion of the methane, it should be possible to alter this geometry to improve methane yields. For example, if the coal-feed nozzle were lowered into stage 2, the backmixing of coal and product gases into the hotter regions of the gasifier would be reduced.

To test this idea, the coal-feed nozzle was extended to various lengths inside stage 2. The best location occurred with the nozzle 1 inch below the stage 1 center line. With Pittsburgh seam coal as feedstock and the nozzle in this position, a methane yield of 25% was obtained for an outlet hydrogen partial pressure of 16 atm. This exceeds the correlation established for the original nozzle position shown in Figure 10 and emphasizes the importance of proper flow patterns in stage 2.

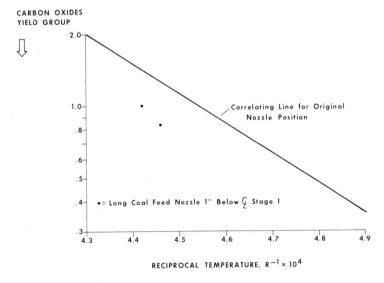

Figure 11. Effect of location of coal feed nozzle on carbon oxides yield group for Pittsburgh seam coal

Carbon oxides yield for this nozzle position was less than expected from earlier correlations, as shown in Figure 11. In this case the product methane and char which were recirculated would have been exposed to less severe temperatures and hence would have contributed less to the yield of carbon oxides. The correlations discussed previously can therefore be viewed as conservative. With proper design to avoid recirculation of product methane into zones of high temperature, the next generation two-stage gasifier can be expected to produce methane yields greater than those obtained in the PEDU.

The PEDU tests discussed here have provided considerable information on the effects of temperature and residence time on the process. We have shown that simple ideas of gasification chemistry and physics can be combined into expressions which are adequate for correlating these data. Extrapolation of these expressions gives estimates for the methane yields which might be obtained from the entrained gasification of coal. For example, the extrapolations of Figure 6 indicate that if residence times could be limited to 2 sec and effective temperatures of 2800°F were attained while maintaining 20 atm hydrogen partial pressure, methane yields of 38% could result. Even at effective temperatures of 2240°F, which are attainable, methane yields of 30% could be obtained if residence times could be limited to 1 or 2 sec.

Whether the exact nature of the rapid, high temperature gasification of coal is properly captured by these correlations to render such extrapolations is valid is, of course, open for discussion. Whether gasifiers with proper geometry can be designed to produce these high yields remains to be seen, but on the basis of the analysis proposed here, the promise is there.

Literature Cited

1. Bituminous Coal Research, "Gas Generator Research and Development—Phase II. Process and Equipment Development," Government Printing Office, Washington, D. C., 1971, 520 pp.
2. Donath, E. E., Glenn, R. A., "Pipeline Gas from Coal by Two-Stage Entrained Gasification," in "Operating Section Proceedings," pp. 65P147-151, American Gas Association, New York, 1965.
3. Bituminous Coal Research, "Gas Generator Research and Development—Phase I. Survey and Evaluation," Government Printing Office, Washington, D. C., 1965, 2 vols.
4. Glenn, R. A., Donath, E. E., Grace, R. J., "Gasification of Coal under Conditions Simulating Stage 2 of the BCR Two-Stage Super-Pressure Gasifier," in "Fuel Gasification," ADVAN. CHEM. SER. (1967) 69, 81–103.
5. Glenn, R. A., Grace, R. J., "An Internally Fired Process Development Unit for Gasification of Coal under Conditions Simulating Stage 2 of the BCR Two-Stage Super-Pressure Process," *Amer. Gas Ass., Synthetic Pipeline Gas Symp.*, 2nd, Pittsburgh, 1968.
6. Glenn, R. A., "Status of the BCR Two-Stage Super-Pressure Process," *Amer. Gas Ass., Synthetic Pipeline Gas Symp.*, 3rd, Pittsburgh, 1970. 18 pp.
7. Grace, R. J., Zahradnik, R. L., "BI-GAS Processing Enters Pilot Plant Stage," *Amer. Gas Ass., Synthetic Pipeline Gas Symp.*, 4th, Chicago, 1972.

8. Grace, R. J., Glenn, R. A., Zahradnik, R. L., "Gasification of Lignite by the BCR Two-Stage Super-Pressure Process," *Amer. Inst. Chem. Eng., Symp. Synthetic Hydrocarbon Fuels Western Coals, Denver, 1970.*
9. Field, M. A., Gill, D. W., Morgan, B. B., Hawksley, P. G. W., "Combustion of Pulverized Coal," British Coal Utilization Research Association, Leatherhead, England, 1967.
10. Moseley, F., Paterson, D., "The Rapid High-Temperature High-Pressure Hydrogenation of Bituminous Coal," *J. Inst. Fuel* (1967) **40**, 523–530.
11. Moseley, F., Paterson, D., "The Rapid High-Temperature Hydrogenation of Coal Chars—Part 1: Hydrogen Pressures up to 100 Atmospheres," *J. Inst. Fuel* (1965) **38**, 13–23.
12. Moseley, F., Paterson, D., "The Rapid High-Temperature Hydrogenation of Coal Chars—Part 2: Hydrogen Pressures up to 1000 Atmospheres," *J. Inst. Fuel* (1965) **38**, 378–391.
13. Zahradnik, R. L., Glenn, R. A., "Direct Methanation of Coal," *Fuel* (1971) **50**, 77–90.
14. Narasimham, K. S., "Flame Length Calculation for Furnace Design," *Erdol Kohle* (1971) **24**, 471–472.
15. Thring, M. W., "The Science of Flames and Furnaces," 2nd ed., Wiley, New York, 1962, 625 pp.

RECEIVED May 25, 1973. Work performed at Bituminous Coal Research, Inc. with support from The Office of Coal Research, Contract No. 14-32-0001-207, and the American Gas Association.

10

Kinetics of Bituminous Coal Char Gasification with Gases Containing Steam and Hydrogen

J. L. JOHNSON

Institute of Gas Technology, 3424 S. State St., Chicago, Ill. 60616

Quantitative correlations developed to describe coal char gasification kinetics are consistent with experimental data obtained under a wide range of conditions with both differential and integral contacting systems. The correlations are developed using data obtained at constant environmental conditions with a thermobalance apparatus and a differential fluid bed system at $1500°-2000°F$ and $1-70$ atm with a variety of gases and gas mixtures. The correlations are based on an idealized model of the gasification process which consists of three consecutively occurring stages: devolatilization, rapid-rate methane formation, and low-rate gasification.

Correlations to define quantitatively the effects of pertinent intensive variables on the kinetics of coal or coal char gasification reactions are necessary for the rational design of commercial systems to convert coal to pipeline gas. The available information which can be applied to the development of such correlations is relatively limited, particularly because the data reported from many studies conducted with integral contacting systems reflect, in part, undefined physical and chemical behavior peculiar to the specific experimental systems used. Although some differential rate data have been obtained with various carbonaceous materials, they cover only narrow ranges of the conditions potentially applicable to commercial gasification systems.

For the last several years the Institute of Gas Technology (IGT) has been conducting a study to obtain fundamental information on the gasification of coals and coal chars; this information could be used with selected literature information to develop engineering correlations which define quantitatively the effects of intensive variables on gasification rates over a wide range of conditions applicable to many gasification processes. The models and correlations developed at present are primarily applicable to bituminous coal chars prepared under mild or severe conditions in either inert or oxidizing atmospheres. Although we have achieved some success in applying these correlations to the gasification of subbituminous and lignite coals for limited conditions, the gasification kinetics of such materials have shown wide deviations from predictions of the correlations at lower temperatures and during initial stages of gasification.

The objectives of this paper are (a) to discuss the models which have been developed, (b) to present the correlations derived from these models, and (c) to demonstrate the consistencies between predictions of these correlations and various experimental gasification data obtained primarily with bituminous coal chars. The experimental information used to develop the models came from two main sources. Initial development of the model applied to the gasification of devolatilized coal char in hydrogen and steam–hydrogen mixtures was based both on data from an IGT study with a high pressure thermobalance apparatus and on differential rate data from the Consolidation Coal Co. on the gasification of Disco char in a small-scale fluid bed (1, 2, 3). The model was extended to the gasification of char containing volatile matter and to gasification with gases containing carbon monoxide, carbon dioxide, and methane as well as steam and hydrogen using data primarily from the thermobalance study.

The thermobalance is particularly useful in obtaining fundamental gasification information because gasification rates can be measured at constant, well defined environmental conditions with it. Most of the information used to formulate the kinetic models developed was based on data from several hundred tests conducted with the thermobalance; this apparatus is described below.

Experimental

The thermobalance is an apparatus capable of continuously weighing a coal sample which is undergoing reaction in a gaseous environment of desired composition at a constant pressure. The temperature can be kept constant or varied (10°F/min is the maximum rate for the apparatus used at IGT). The nature of gas–solid contact with the apparatus used in this study is shown in Figure 1. The coal sample is contained in the annular space of a wire mesh basket bounded on the inside by a hollow, stainless steel tube and on the outside by a wire mesh screen. To facilitate mass and heat transfer between the bed and its environment, the thickness of the bed is only 2-3 particle diameters when using $-20+40$ US sieve-size particles. Gas flow rates used with this system are sufficiently large relative to gasification rates so that gas conversion is limited to less than 1% for devolatilized coal char.

In a typical test the wire mesh basket is initially in an upper, cooled portion of the reactor in which a downward, inert gas flow is maintained. During this time the desired temperature and pressure conditions are established in a lower, heated portion of the reactor in the presence of a flowing gas. A test is initiated by lowering the basket into the heated reaction zone, a procedure which takes 5-6 sec. Theoretical computation shows that about 2 min are needed for the sample to achieve reactor temperature as measured by several thermocouples surrounding the basket in the reaction zone. This computation is reasonably corroborated by various kinetic indications and by the behavior of the thermocouples in re-attaining their preset temperatures. The sample is kept in the heated portion of the reactor for the specified time while its weight is continuously recorded. The test is terminated by raising the basket back to the upper, cooled portion of the reactor.

During a test, the dry feed gas flow rates are measured by an orifice meter and the dry product gas flow rates measured by a wet-test meter.

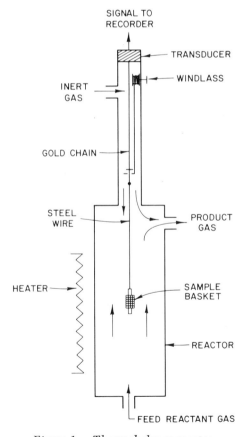

Figure 1. Thermobalance reactor

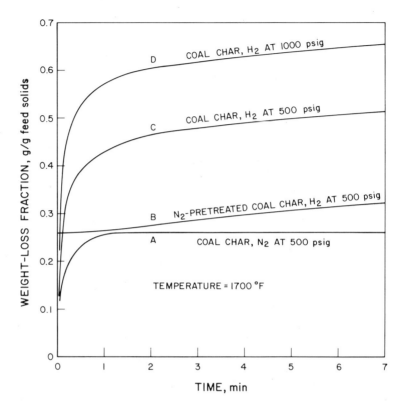

Figure 2. Typical weight-loss curves obtained with high pressure thermobalance for gasification in hydrogen and in nitrogen

Periodic samples of product gas are taken to determine the composition by mass spectrometer. Feed and product steam flow rates are measured gravimetrically, and the solids residues are analyzed for total carbon and hydrogen.

Figure 2 shows typical, smoothed, weight loss-*vs.*-time characteristics obtained using an air-pretreated hvab Pittsburgh coal char from the Ireland mine. These curves are discussed in more detail later. The composition of the coal char, used extensively in the experimental study, is given in Table I.

Kinetic Models

When a coal or coal char containing volatile matter is initially subjected to an elevated temperature, a series of complex physical and chemical changes occur in the coal's structure, accompanied by thermal pyrolysis reactions which result in devolatilization of certain coal components. The distribution of the evolved products of the reactions, which initiate at less than 700°F and can be considered to occur almost instantaneously at temperatures greater than 1300°F, is generally a function of the temperature, pressure, and gas composition existing during devolatilization and of the subsequent thermal and environmental history of the gaseous phase (including entrained liquids) prior to quenching.

Table I. Composition of Air-Pretreated hvab Pittsburgh No. 8 Coal Char (Ireland Mine)

Ultimate Analysis, dry	wt %
Carbon	71.1
Hydrogen	4.26
Oxygen	8.85
Nitrogen	1.26
Sulfur	3.64
Ash	10.89
Total	100.00
Proximate Analysis, dry	
Fixed carbon	60.7
Volatile	28.4
Ash	10.9
Total	100.0

When devolatilization occurs in the presence of a gas containing hydrogen at an elevated pressure, in addition to thermal pyrolysis reactions, coals or coal chars containing volatile matter also exhibit a high although transient reactivity for methane formation. Although some investigators have suggested that this process occurs simultaneously with thermal pyrolysis reactions, studies done with a greater time resolution indicated that this rapid-rate methane formation occurs at a rate which is at least an order of magnitude slower than devolatilization (4, 5). In this sense it occurs after devolatilization.

The amount of carbon gasified to methane during the transient high reactivity increases significantly with increases in hydrogen partial pressure (4, 5, 6). Experimental evidence indicates that at sufficiently high hydrogen partial pressures virtually all of the carbon not evolved during devolatilization can be gasified quickly to methane by this process (6). This is contrary to some proposed models which assume that only a limited amount of carbon can be gasified in this reaction stage regardless of the hydrogen partial pressure (7, 8).

At temperatures greater than 1700°F the transient reactivity for rapid rate methane formation exists only briefly. For coals or coal chars prepared in inert atmospheres this period is seconds or less (6). IGT's studies suggest that for air-pretreated coal chars, this period is more extended although the total amounts of carbon which can be gasified by this process at a given temperature and hydrogen partial pressure are comparable for coals and coal chars prepared at sufficiently low temperatures in either inert gas or air.

After the devolatilization and rapid-rate methane formation stages are completed, char gasification occurs at a relatively slow rate; various models to describe the gasification kinetics of this material for various limited ranges of conditions have been proposed. The differential rates of reaction of devolatilized coal chars are a function of temperature, pressure, gas composition, carbon conversion, and prior history.

General Assumptions in the Development of Models. The models developed in this study for the quantitative description of coal char gasification kinetics assume that the overall gasification occurs in three consecutive stages: (1) devolatilization, (2) rapid-rate methane formation, and (3) low-rate gasification. The reactions in these stages are independent. Further, a feed coal char contains two types of carbon—volatile carbon and base carbon. Volatile carbon can be evolved solely by thermal pyrolysis, independent of the gaseous medium. The distribution of evolved products derived from the volatile carbon fraction is not defined in the model. In any application of the model to an integral contacting system, the devolatilization products are estimated by extrapolation or interpolation of yields obtained in pilot-scale fluid or moving-bed systems. This procedure can be used for narrow ranges of conditions and only for very similar contacting systems. Base carbon remains in the coal char after devolatilization is complete. This carbon can be subsequently gasified in either the rapid-rate methane formation stage or the low-rate gasification stage.

Initial amounts of volatile and base carbon are estimated from standard analyses of the feed coal char:

C_v (volatile carbon), grams/gram feed coal

$= C_t^\circ$ (total carbon), grams/gram feed coal

$- C_b^\circ$ (base carbon), grams/gram feed coal   (1)

where C_t° represents the total carbon in the feed coal char obtained from an ultimate analysis, and C_b° represents the carbon in the fixed carbon fraction of the feed coal as determined in a proximate analysis. Note that C_b° does not equal the fixed carbon fraction because the fixed carbon fraction includes, in addition to carbon, other organic coal components not evolved during standard devolatilization.

Experimental results from thermobalance or free-fall tests conducted at IGT indicate that the assumption of a constant volatile carbon fraction is valid for coal heat-up rates as high as 200°F/sec. However, other studies conducted with extremely rapid heat-up rates (10^4 to 10^{7}°F/sec) have shown that quantities of carbon evolved during thermal pyrolysis can exceed the volatile carbon fraction defined in this model (9, 10). An allowance for the increase in evolved carbon would, therefore, have to be made in systems using such high heating rates.

The base carbon conversion fraction, X, is defined as:

$$X = \frac{\text{base carbon gasified}}{\text{base carbon in feed coal char}} = \frac{C_b{}^\circ - C_b}{C_b{}^\circ} \qquad (2)$$

where C_b = base carbon in coal char at an intermediate level of gasification, grams/gram feed coal char.

When making a kinetic analysis of the thermobalance data it was necessary to relate the measured values of weight-loss fraction, $\Delta W/W_o$, to the base carbon conversion fraction, X. When devolatilization is complete, essentially all of the organic oxygen has been gasified. Thereafter, additional weight loss, which primarily results from gasification of the base carbon, is accompanied by the evolution of a constant fraction of noncarbon components in the coal such as nitrogen, hydrogen, and sulfur. Estimates of an average value of the fraction of noncarbon components gasified along with the base carbon for each type of coal char tested have been based on analyses of char residues obtained in thermobalance and pilot-scale fluid-bed tests. With this simplifying assumption, the following relationships result:

$$C_b{}^\circ = (1 - V - A)(1 - \gamma) \qquad (3)$$

and

$$C_b{}^\circ = C_b - [(\Delta W/W_0) - V](1 - \gamma) \qquad (4)$$

where

V = volatile matter in feed coal char (including moisture), grams/gram feed coal char

A = nongasifiable matter in feed coal char (including ash and some sulfur), grams/gram feed coal char

γ = noncarbon matter evolved along with base carbon, grams/gram base carbon evolved

Thus, from Equations 2, 3, and 4

$$X = \frac{(\Delta W/W_0) - V}{1 - V - A} \qquad (5)$$

The results in Figure 2 can be interpreted by the three reaction stages defined above. Curve A corresponds to a test in which air-pretreated Ireland mine coal char was exposed to a nitrogen atmosphere at 500 psig. During the first minute when the sample is heating in the thermobalance, the weight loss corresponds to the evolution of volatile matter. After this period no further significant weight loss occurs. The total weight loss of *ca.* 26% corresponds closely to the volatile matter in the feed coal char obtained by standard proximate analysis. Within the context of the three reaction stages defined, weight loss in this test occurs entirely in the devolatilization stage where all of the volatile carbon has been gasified; all of the base carbon remains in the devolatilized coal char. When the char resulting from this test is then exposed to hydrogen at 500 psig (curve B), the reaction of the hydrogen with base carbon to form methane results in further weight loss. This reaction, which takes place at a much slower rate than devolatilization, occurs in the low-rate gasification stage. With this particular sample there was no reaction in the rapid-rate methane formation stage because the reactivity of the coal char in this stage was destroyed during prolonged exposure to nitrogen.

Weight loss as a result of reaction during rapid-rate methane formation is illustrated by curve C where a sample of the original coal char was exposed only to hydrogen at 500 psig with no initial exposure to nitrogen. The weight loss during the first minute of this test is considerably greater than the corresponding weight loss during this period when the coal char was exposed to nitrogen (curve A). The difference in the weight loss between curves C and A during the first minute is caused by gasification of base carbon in the rapid-rate methane formation stage. Weight losses of the amount shown by curve B during this initial period are negligible. This is consistent with the assumption that base carbon gasification in the rapid-rate methane formation stage and in the low-rate gasification stage occurs consecutively. Curve D is qualitatively similar to curve C except that there is a greater weight loss from rapid-rate methane formation resulting from the higher hydrogen pressure.

Correlations for Rapid-Rate Methane Formation Stage. The amount of base carbon gasified during the rapid-rate methane formation stage can be estimated by the base carbon conversion level, X_R, obtained from weight loss-*vs.*-time characteristics 2 min after a sample is lowered into the reactor. As indicated previously, this corresponds to the time required for coal heat-up. During this period negligible conversion occurs in the low-rate gasification stage although devolatilization and rapid-rate methane formation reactions should be complete at temperatures above *ca.* 1500°F. Values of X_R have been correlated with hydrogen partial pressure, P_{H_2}, according to the following expression for data obtained in tests conducted at varied conditions:

$$M(X_R) = \int_0^{X_R} \frac{\exp(+\alpha X^2)\mathrm{d}X}{(1-X)^{2/3}} = 0.0092 f_R P_{H_2} \qquad (6)$$

where

P_{H_2} = hydrogen partial pressure, atm
f_R = relative reactivity factor for rapid-rate methane formation dependent on the particular carbonaceous solid (defined as unity for air-pretreated Ireland mine coal char)
α = kinetic parameter dependent on gas composition and pressure

The value of α in the above expression is ca. 0.97 for tests done in pure hydrogen or in hydrogen–methane mixtures and is approximately equal to 1.7 for a variety of gas compositions containing steam and hydrogen. This parameter is discussed in greater detail in a later section on the low-rate gasification stage. However, for the case where $\alpha = 0.97$, then $M(X_R) \cong -\ln(1-X_R)$.

Figure 3 is a plot of the function $M(X_R)$ vs. hydrogen partial pressure, P_{H_2}, for tests conducted on the thermobalance with air-pretreated Ireland mine coal char; note the good agreement between these data and

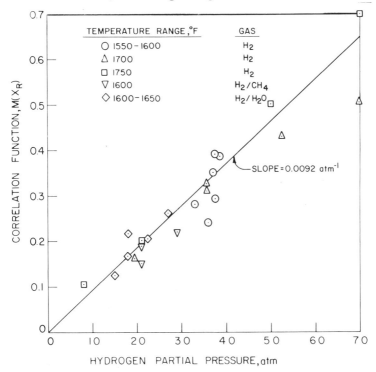

Figure 3. Correlation of base carbon conversion for gasification in the rapid-rate methane formation stage (thermobalance data)

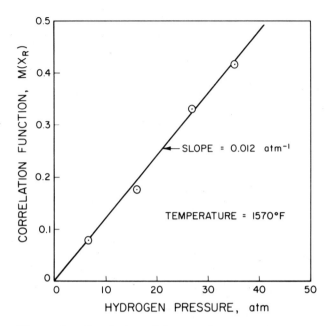

Figure 4. Correlation of base carbon conversion for gasification in a fluid bed in the rapid-rate methane formation stage

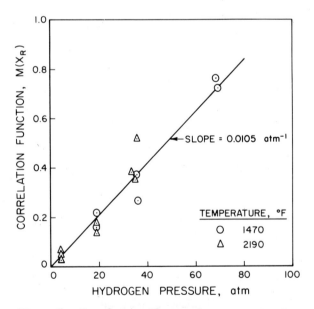

Figure 5. Correlation of base carbon conversion for gasification in a fixed bed in the rapid-rate methane formation stage

the correlation form given in Equation 6. Equation 6 also appears to be reasonably applicable to the gasification of some coals as well as to coal chars. Data obtained by Birch et al. (11) for the hydrogenation of Brown coal in a fluid bed and by Hiteshue et al. (12) for the hydrogenation of a hvab Pittsburgh coal in a fixed bed are given in Figures 4 and 5. The procedures used to treat the data from these two investigations have been described (13). Relatively small variations in f_R values are exhibited by the different materials. Similar small degrees of variation have also been noted for other bituminous coal chars tested at IGT using the thermobalance.

The gasification of base carbon in the rapid-rate methane formation stage depends apparently only on hydrogen partial pressure and not on the partial pressures of other gaseous species normally present in gasifying atmospheres. This is partly indicated in Figure 3 for tests conducted with hydrogen–methane and hydrogen–steam mixtures and has also been observed with synthesis-gas mixtures. In a system containing no hydrogen, base carbon is not evolved except through gasification in the low-rate gasification stage. In Figure 6 this effect is illustrated for tests conducted

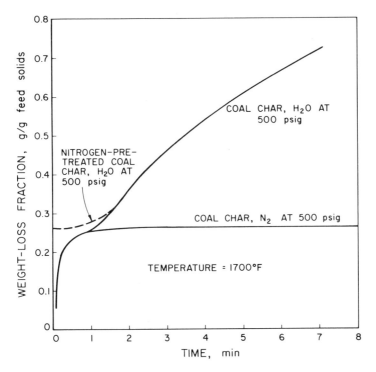

Figure 6. Weight loss curves obtained in high pressure thermobalance for gasification in steam and in nitrogen

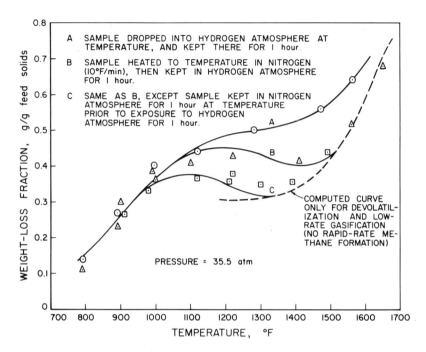

Figure 7. *Effect of prior exposure to nitrogen on gasification in the rapid-rate methane formation stage*

with pure steam. After the first few minutes there is no difference in the weight-loss traces for a sample of air-pretreated coal char initially lowered into a steam atmosphere and one which was devolatilized in nitrogen prior to being exposed to the steam atmosphere. This behavior can be compared with results shown in Figure 2 for similar tests conducted in hydrogen.

Although the effects of pretreatment temperature on reactivity in the rapid-rate methane formation stage have not been studied systematically in this investigation, some indication is apparent from the results in Figure 7. It shows total weight losses for materials subjected to different pretreatments in nitrogen obtained after gasification of air-pretreated coal char in hydrogen for 1 hr. Below 1000°F no effect of the nitrogen pretreatment is apparent on subsequent weight loss in hydrogen. Above this temperature, however, the total weight loss for materials initially exposed to nitrogen tends to decrease with increasing temperature; above *ca.* 1500°F no rapid-rate methane formation occurs with these materials.

The correlation described by Equation 6 is based on data obtained from thermobalance tests conducted above 1500°F where coal char samples were submitted to specific heat-up rates characteristic of the experimental apparatus (approximately 30°F/sec). The fact that this correlation seems to apply for data obtained in other experimental systems where heat-up rates as high as 200°F/sec were used suggests that within a limited range of heat-up rates, base carbon conversion in the rapid-rate methane formation stage is independent of heat-up rate for final temperatures greater than 1500°F. This conclusion applies only when reaction in the rapid-rate methane formation stage goes to completion. At sufficiently low temperatures, the amount of base carbon conversion which can be attributed to rapid-rate methane formation is less than that which would be predicted by Equation 6, even for exposure to hydrogen for as long as 1 hr (*13*).

Data obtained in experiments done at low temperatures, such as those in Figure 7, have been correlated using a more detailed model to describe the rapid-rate methane formation process prior to the completion of this reaction. This model is described in a previous publication (*13*). Certain characteristics of this model rationalize the independence of total base carbon conversion in the rapid-rate methane formation stage from heating rate and final temperature for tests done above 1500°F. The following critical steps assumed in the model relate to this range of conditions:

$$A_0 \rightarrow A_* \tag{7}$$

$$A_* \rightarrow B \tag{8}$$

$$\text{base carbon} + \text{hydrogen} \xrightarrow{A_*} \text{methane} \tag{9}$$

This model, which is qualitatively similar to one proposed by Mosely and Patterson (*6*), assumes that the coal char initially forms an active intermediate, A_*, (Equation 7) which catalyzes the reaction between base carbon and gaseous hydrogen to form methane (Equation 9). This reaction, however, competes with a reaction in which A_* deactivates to form the inactive species, B (Equation 8).

The following expression is assumed to describe the rate of reaction in Equation 9:

$$dX/dt = f_R k_3(T) P_{H_2} (1 - X)^{2/3} \exp(-\alpha X^2) N_{A_*} \tag{10}$$

where

$k_3(T)$ = rate constant dependent on temperature, T

N_{A_*} = concentration of species, A, mole/mole of base carbon

t = time

The dependence of the conversion rate, dX/dt, on conversion fraction, X, shown in Equation 10 is the same as that used in correlations presented in a later section which were developed to describe gasification in the low-rate gasification stage. With the models assumed, the term $(1 - X)^{2/3}$ is proportional to the effective surface area undergoing gasification, and the term $\exp(-\alpha X^2)$ represents the relative reactivity of the effective surface area which decreases with increasing conversion level for positive values of α.

The reaction rates in Equations 7 and 8 are assumed to be first order, leading to the expression:

$$d(N_{A_*} + N_{A_0})/dt = - k_2(T) \cdot N_{A_*} \quad (11)$$

where

$k_2(T)$ = rate constant, dependent on temperature

N_{A_0} = concentration of species, A_0, mole/mole base carbon

Combining Equations 10 and 11 leads to:

$$\frac{dX}{d(N_{A_*} + N_{A_0})} = - f_R \frac{k_3(T)}{k_2(T)} P_{H_2}(1 - X)^{2/3} \exp(-\alpha X^2) \quad (12)$$

If it is assumed that $k_3(T)/k_2(T)$ is independent of temperature and is equal to β, Equation 12 can be integrated to yield the following expression for the condition at which all of A_0 has been converted to B.

$$M(X_R) = \int_0^{X_R} \frac{\exp(\alpha X^2)dX}{(1 - X)^{2/3}]} = f_R \beta \cdot N^0{}_{A_0} P_{H_2} \quad (13)$$

A comparison of this expression with Equation 6 indicates that $\beta N^0{}_{A_0}$ = 0.0092 atm^{-1}. Since no definition of the temperature history was required to develop Equation 13, the suggested model indicates that the amount of base carbon conversion to methane during the rapid-rate methane formation step is independent of heat-up rate or temperature level when the intermediate, A, has been completely deactivated.

Correlations for Low-Rate Gasification Stage. For practical purposes coal chars undergo low-rate gasification only after the devolatilization and rapid-rate methane formation reactions are completed. Results obtained with the thermobalance indicate that at greater than 1500°F char reactivity over a major range of carbon conversion in the low-rate stage is substantially the same whether devolatilization occurs in nitrogen or in a gasifying atmosphere under the same conditions. Therefore, this study treats low-rate char gasification as a process essentially independent of devolatilization conditions with one important exception; the temperature of devolatilization. It has been shown in this study as well as by Blackwood (14) that the reactivity of a char at a given temperature, T, decreases with increasing pretreatment temperature, T_o, when $T_o > T$. This effect is quantitatively represented in the correlations below. However, the model adopted does not account for pretreatment effects on gasification during initial stages of char gasification which occur particularly at gasification temperatures less than 1600°F. At these lower tem-

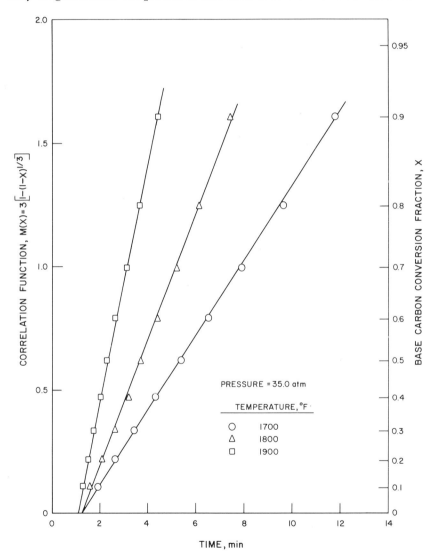

Figure 8. Effect of temperature on low-rate gasification in steam (thermobalance data)

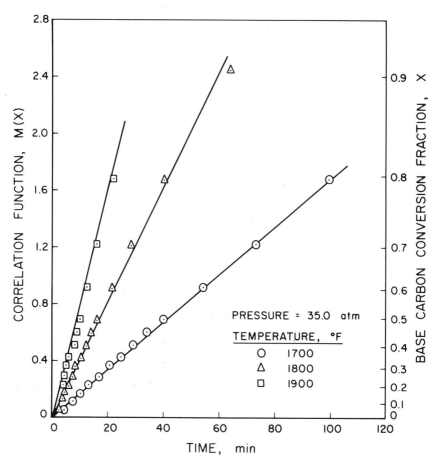

Figure 9. Effect of temperature on low-rate gasification in hydrogen (thermobalance data)

peratures, specific pretreatment conditions such as gas atmosphere and time of pretreatment produce complex effects during subsequent gasification for base carbon conversions of less than 10% (15). These limitations have no practical significance in using the simplified model developed to describe coal char gasification kinetics at higher temperatures or for base carbon conversion levels sufficiently greater than 10%.

The gasification data of Zielke and Gorin (2, 3) and Goring et al. (1) for fluid-bed gasification of Disco char as well as the bulk of data obtained in IGT studies with the high pressure thermobalance and pilot-scale fluid beds were used to evaluate parameters in a quantitative model developed to describe coal char gasification kinetics over a wide range of conditions in the low-rate gasification stage. Three basic reactions were assumed to occur in gases containing steam and hydrogen:

Reaction I: $H_2O + C \rightleftarrows CO + H_2$

Reaction II: $2H_2 + C \rightleftarrows CH_4$

Reaction III: $H_2 + H_2O + 2C \rightleftarrows CO + CH_4$

Reaction I is the conventional steam–carbon reaction which is the only one that occurs in pure steam at elevated pressures or with gases

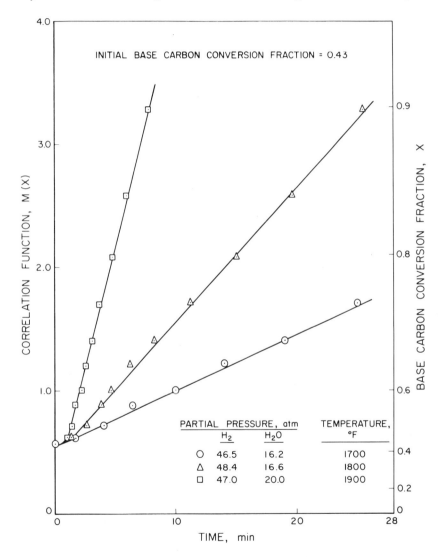

Figure 10. *Effect of temperature on low-rate gasification in steam–hydrogen mixtures (thermobalance data)*

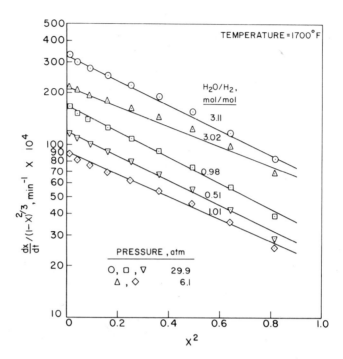

Figure 11. Effects of pressure and gas composition on low-rate gasification in steam–hydrogen mixtures (1, 3)

containing steam at low pressure. Although at elevated temperatures this reaction is affected by thermodynamic reversibility only for relatively high steam conversions, the reaction is severely inhibited by the poisoning effects of hydrogen and carbon monoxide at steam conversions far removed from equilibrium for this reaction. Some investigators have also noted inhibition by methane. (Although some methane has been detected in gaseous reaction products when gasification is conducted with pure steam, it is uncertain whether this methane results from the direct reaction of steam with carbon or from the secondary reaction of hydrogen, produced from the steam–carbon reaction, with the carbon in the char.)

Reaction II, the only reaction that could occur in pure hydrogen or in hydrogen–methane mixtures, depends greatly on the hydrogen partial pressure. Many investigators have found that at elevated pressures its rate is directly proportional to the hydrogen partial pressure.

The stoichiometry of Reaction III limits its occurrence to systems in which both steam and hydrogen are present. Although this reaction is the stoichiometric sum of Reactions I and II, this model considers it to be a third, independent gasification reaction. Reaction III, arbitrarily assumed to occur in the development of this model to facilitate correlation of experimental data, has been suggested by Blackwood and McGrory (16) as being necessary in such a system. Curran and Gorin (17) also assumed this reaction to correlate kinetic data for gasification of lignite at 1500°F in steam-hydrogen-containing gases.

The correlations developed in this study to describe kinetics in the low-rate gasification stage are summarized as follows:

$$dX/dt = f_L k_T (1 - X)^{2/3} \exp(-\alpha X^2) \quad (14)$$

where

$$k_T = k_I + k_{II} + k_{III} \quad (15)$$

Here, k_I, k_{II}, and k_{III} are rate constants for the individual reactions considered. It is assumed that each of the three reactions occurs independently but that the rate of each is proportional to the same surface area term, $(1 - X)^{2/3}$ and surface reactivity term, $\exp(-\alpha X^2)$.

Individual parameters in Equations 14 and 15 are defined as functions of temperature and pressure according to:

$$f_L = f_0 \exp(8467/T_0) \quad (16)$$

$$k_I = \frac{\exp(9.0201 - 31{,}705/T)\left(1 - \dfrac{P_{CO}P_{H_2}}{P_{H_2O}K^E_I}\right)}{\left[1 + \exp(-22.2160 + 44{,}787/T)\left(\dfrac{1}{P_{H_2O}} + 16.35\dfrac{P_{H_2}}{P_{H_2O}} + 43.5\dfrac{P_{CO}}{P_{H_2O}}\right)\right]^2} \quad (17)$$

$$k_{II} = \frac{P^2_{H_2}\exp(2.6741 - 33{,}076/T)\left(1 - \dfrac{P_{CH_4}}{P^2_{H_2}K^E_{II}}\right)}{[1 + P_{H_2}\exp(-10.4520 + 19{,}976/T)]} \quad (18)$$

$$k_{III} = \frac{P^{1/2}_{H_2}P_{H_2O}\exp(12.4463 - 44{,}544/T)\left(1 - \dfrac{P_{CH_4}P_{CO}}{P_{H_2}P_{H_2O}K^E_{III}}\right)}{\left[1 + \exp(-6.6696 + 15{,}198/T)\left(P^{1/2}_{H_2} + 0.85 P_{CO} + 18.62\dfrac{P_{CH_4}}{P_{H_2}}\right)\right]^2} \quad (19)$$

$$\alpha = \frac{52.7 P_{H_2}}{1 + 54.3 P_{H_2}} + \frac{0.521 P^{1/2}_{H_2} P_{H_2O}}{1 + 0.707 P_{H_2O} + 0.50 P_{H_2}^{1/2} P_{H_2O}} \quad (20)$$

where

Table II. Experimental and Calculated Rate Constants for Gasification of Air-Pretreated Ireland Mine Coal Char in Hydrogen

Temp, °F	H_2 Pressure, atm	Base Carbon Conversion Range, Fraction	Rate Constant, fLT, min^{-1} Experimental	Calculated
1600	36.5	0-0.314	0.0068	0.0069
1650	36.3	0-0.541	0.0117	0.0112
1650	36.7	0-0.592	0.0111	0.0113
1650	36.7	0-0.654	0.0097	0.0113
1700	19.3	0-0.197	0.0106	0.0090
1700	36.2	0-0.329	0.0167	0.0181
1700	53.2	0-0.449	0.0276	0.0273
1700	69.9	0-0.508	0.0292	0.0365
1700[a]	35.1	0-0.802	0.0175	0.0175
1750	35.4	0-0.348	0.0261	0.0277
1750	35.4	0-0.509	0.0274	0.0277
1750	36.1	0-0.640	0.0237	0.0283
1750	35.2	0-0.367	0.0272	0.0276
1770[a]	1.0	0	0	0.0002
1770[a]	18.1	0-0.069	0.0175	0.0152
1770[a]	36.6	0-0.134	0.0350	0.0341
1770[a]	36.1	0-0.570	0.0290	0.0335
1770[a]	52.9	0-0.180	0.0500	0.0510
1770[a]	69.9	0-0.250	0.0700	0.0686
1800	21.6	0-0.263	0.0222	0.0239
1800	49.9	0-0.263	0.0641	0.0617
1800[a]	36.0	0-0.910	0.0416	0.0414
1900[a]	35.3	0-0.850	0.0910	0.0920

[a] In these tests, the air-pretreated coal char was either initially devolatilized in nitrogen for 1 hr at the temperature subsequently used for gasification in a gasifying medium or it was devolatilized and partially gasified in an integral fluid-bed test using a steam–hydrogen feed gas.

$K^E_I, K^E_{II}, K^E_{III}$ = equilibrium constants for Reactions I, II, and III, considering carbon as graphite

T = reaction temperature, °R

T_o = maximum temperature to which char has been exposed prior to gasification, °R (if $T_o < T$, then a value of $T_o = T$ is used in Equation 16)

$P_{H_2}, P_{H_2O}, P_{CO}, P_{CH_4}$ = partial pressures of H_2, H_2O, CO, and CH_4, atm

f_o = relative reactivity factor for low-rate gasification which depends on the particular carbonaceous solid

Values of f_o obtained in this study were based on the definition $f_o = 1.0$ for a specific batch of air-pretreated Ireland mine coal char. Samples of this coal char obtained from different air-pretreatment tests exhibited some variations in reactivity as determined by thermobalance tests conducted at standard conditions. The values of f_o so determined ranged from approximately 0.88 to 1.05. Results of tests made with the thermobalance, using a variety of coals and coal chars, have indicated that the relative reactivity factor, f_o, generally tends to increase with decreasing rank although individual exceptions to this trend exist. Values have been obtained which range from 0.3 for a low-volatile bituminous coal char to about 10 for a North Dakota lignite. The reactivity of the Disco char used in gasification studies conducted by the Consolidation Coal Co. (1, 2, 3) is $f_o = 0.488$.

Table III. Experimental and Calculated Rate Constants for Gasification of Air-Pretreated Ireland Mine Coal Char in Steam and Steam-Hydrogen Mixtures

Temp, °F	Partial Pressure, atm		Base Carbon Conversion Range, Fraction	Rate Constant, $f_L k_T$, min^{-1}	
	H_2	H_2O		Experimental	Calculated
1700	—	35.0	0-0.941	0.1547	0.1736
1700	27.3	29.8	0-0.977	0.0598	0.0528
1700	12.8	23.6	0-0.327	0.0706	0.0566
1700	17.8	18.4	0-0.272	0.0365	0.0385
1700	19.2	17.5	0-0.651	0.0356	0.0361
1700	18.2	17.9	0-0.816	0.0344	0.0373
1700	23.3	12.4	0-0.300	0.0339	0.0274
1700	23.2	12.5	0-0.429	0.0262	0.0275
1700	23.8	11.9	0-0.511	0.0190	0.0267
1700	25.0	11.3	0-0.725	0.0224	0.0261
1700[a]	13.7	48.5	0.43-0.908	0.0831	0.1110
1700[a]	46.5	16.2	0.43-0.781	0.0446	0.0385
1700[a]	28.3	32.5	0.43-0.832	0.0425	0.0564
1750	17.6	18.5	0-0.801	0.0526	0.0763
1750	18.3	18.2	0-0.495	0.0613	0.0741
1750	17.6	18.5	0-0.702	0.0667	0.0763
1750	18.0	18.5	0-0.940	0.0917	0.0758
1750	5.3	30.8	0-0.573	0.1803	0.2162
1750	23.2	13.2	0-0.320	0.0441	0.0532
1750	26.9	9.2	0-0.334	0.0406	0.0422
1750	17.6	18.3	0-0.350	0.0670	0.0815
1800	—	35.0	0-0.990	0.2559	0.2815
1800	48.4	16.6	0-0.956	0.1115	0.1130
1800	32.4	27.6	0-0.981	0.1643	0.1710
1900	—	35.0	0-0.992	0.4795	0.4366
1900[a]	17.2	42.7	0.43-0.996	0.6570	0.8750
1900[a]	47.0	20.0	0.43-0.999	0.4000	0.3900
1900[a]	33.2	33.2	0.43-0.997	0.680	0.607

[a] See Table II.

Table IV. Experimental and Calculated Rate Constants for Gasification

Temp, °F	Partial Pressure, atm				
	CO	CO_2	H_2	H_2O	CH_4
1550	0.17	0.32	5.92	28.11	1.68
1600	0.95	0.28	14.46	11.07	10.73
1600	0.65	0.35	12.74	13.82	9.68
1600	0.65	0.38	14.49	11.33	10.29
1600	0.17	0.17	7.73	14.66	9.65
1600	0.10	0.33	6.41	26.58	2.18
1600	8.60	6.38	11.07	7.50	2.38
1600	7.57	1.70	18.87	6.84	1.04
1600	2.28	0.46	19.03	6.57	8.48
1650	0.55	0.26	11.76	14.66	9.65
1650	0.10	0.29	6.27	26.47	2.51
1650	2.24	3.25	6.00	23.02	1.29
1700	1.85	1.23	8.47	24.71	0.04
1700	6.05	1.64	15.81	12.91	0.16
1700	5.27	4.17	10.43	15.87	0.05
1700	4.52	3.79	9.54	18.11	0.12
1700	6.90	4.45	10.88	13.58	0.14
1700	2.26	3.12	3.49	28.07	0.11
1700	4.33	4.69	4.18	22.54	0.07
1700[a]	0.60	2.80	4.30	60.10	0.40
1700[a]	5.0	9.4	19.8	25.8	2.1
1700[a]	1.8	3.1	6.3	19.2	0.4
1700[a]	5.9	10.8	13.2	28.0	0.9
1700[a]	4.3	5.1	6.7	16.8	0.5
1750	6.07	4.62	9.69	13.47	2.34
1750	3.31	4.80	8.12	19.08	1.23
1750	2.49	3.94	4.99	24.18	0.94
1750	5.87	6.61	7.03	16.00	0.16
1750	6.6	4.9	10.6	13.9	0.3
1800[a]	6.2	7.4	18.1	31.6	0.3
1800[a]	1.8	4.9	11.6	47.7	0.2
1800[a]	10.7	6.7	17.7	19.1	1.5
1800[a]	4.5	2.6	7.4	13.3	0.3
1800[a]	6.9	3.4	10.7	8.4	0.4
1800[a]	6.0	9.7	13.4	30.5	0.4
1800[a]	2.9	11.1	14.1	36.5	0.4
1800[a]	5.9	4.9	7.6	14.1	0.2
1800[a]	1.1	3.3	4.3	24.7	0.2
1800[a]	12.9	4.9	7.5	9.4	0.3
1800[a]	0.5	2.0	4.0	27.4	0.1
1800[a]	—	1.2	2.6	30.1	0.1
1900[a]	4.0	7.4	18.0	35.7	0.2
1900[a]	3.2	5.3	11.8	42.9	0.1
1900[a]	12.3	7.8	23.7	16.0	1.3

of Air-Pretreated Ireland Mine Coal Char in Synthesis Gas

Base Carbon Conversion Range, Fraction	Rate Constant, $f_L k_T$, min^{-1}	
	Experimental	Calculated
0-0.277	0.0045	0.0034
0-0.130	0.0002	0.0006
0-0.143	0.0007	0.0009
0-0.159	0.0006	0.0006
0-0.228	0.0022	0.0044
0-0.571	0.0146	0.0093
0-0.106	0.0001	0.0001
0-0.258	0.0022	0.0033
0-0.176	0.0004	0.0002
0-0.228	0.0026	0.0036
0-0.703	0.0234	0.0233
0-0.426	0.0089	0.0099
0-0.745	0.0272	0.0400
0-0.550	0.0122	0.0122
0-0.576	0.0143	0.0128
0-0.619	0.0166	0.0161
0-0.543	0.0125	0.0096
0-0.817	0.0516	0.0481
0-0.736	0.02680	0.0216
0.43-0.871	0.104	0.128
0.43-0.663	0.0200	0.0169
0.43-0.659	0.0188	0.0274
0.43-0.661	0.0136	0.0172
0.43-0.556	0.0110	0.0117
0-0.452	0.0092	0.0095
0-0.842	0.0404	0.0315
0-0.952	0.0771	0.0588
0-0.697	0.0230	0.0217
0-0.623	0.0170	0.0175
0.43-0.933	0.0880	0.0880
0.43-0.979	0.236	0.248
0.43-0.776	0.043	0.0300
0.43-0.784	0.0395	0.0448
0.43-0.678	0.0299	0.0193
0.43-0.878	0.0736	0.0849
0.43-0.899	0.105	0.146
0.43-0.771	0.0431	0.0389
0.43-0.959	0.182	0.199
0.43-0.647	0.0111	0.0101
0.43-0.892	0.246	0.268
0.43-0.914	0.273	0.271
0.43-0.998	0.391	0.395
0.43-0.991	0.579	0.513
0.43-0.845	0.0820	0.0964

Table IV.

Temp, °F	Partial Pressure, atm				
	CO	CO$_2$	H$_2$	H$_2$O	CH$_4$
1900[a]	3.9	2.9	8.8	11.9	0.1
1900[a]	7.1	·3.3	11.4	7.9	0.2
1900[a]	1.5	3.1	6.6	20.7	0.2
1900[a]	8.9	10.5	15.2	28.2	0.5
1900[a]	19.2	11.6	16.6	15.5	0.7
1900[a]	3.8	7.8	11.0	41.0	0.3
1900[a]	6.1	5.3	8.3	13.5	0.1
1900[a]	1.4	3.7	4.7	23.7	0.1
1900[a]	10.6	5.2	8.4	10.5	0.1
1900[a]	0.3	1.3	2.7	29.7	—
1900[a]	1.0	3.3	7.0	56.7	0.1
2000[a]	3.2	5.3	13.4	46.1	0.2
2000[a]	7.5	6.0	18.6	35.8	0.5
2000[a]	0.8	0.3	33.9	32.5	0.5

[a] See Table II.

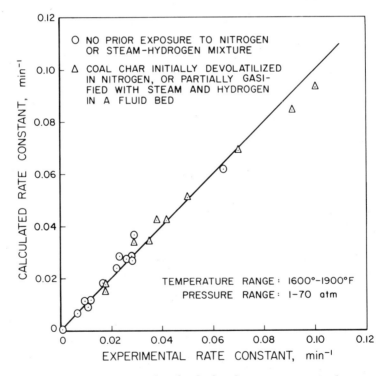

Figure 12. *Experimental and calculated rate constants for low-rate gasification in hydrogen (thermobalance data)*

Continued

Base Carbon Conversion Range, Fraction	Rate Constant, $f_L k_T$, min^{-1}	
	Experimental	Calculated
0.43-0.956	0.141	0.169
0.43-0.840	0.0638	0.0713
0.43-0.981	0.364	0.415
0.43-0.999	0.227	0.200
0.43-0.824	0.0721	0.0609
0.43-1.009	0.438	0.442
0.43-0.984	0.191	0.142
0.43-1.033	0.500	0.471
0.43-0.872	0.085	0.064
0.43-1.028	0.655	0.817
0.43-1.001	1.000	0.892
0.43-0.991	1.36	1.27
0.43-1.047	0.771	0.873
0.43-0.987	1.462	1.297

An integrated form of Equation 14 was used to evaluate certain parameters in the above correlations, based on data obtained with the thermobalance.

$$M(X) = \int_0^X \frac{\exp(+\alpha X^2)dX}{(1-X)^{2/3}} = \int_0^{X_R} \frac{\exp(+\alpha X^2)dX}{(1-X)^{2/3}} \quad (21)$$

$$+ \int_{X_R}^X \frac{\exp(-\alpha X^2)dX}{(1-X)^{2/3}} = \int_0^{X_R} \frac{\exp(+\alpha X^2)dX}{(1-X)^{2/3}} + f_L k_T t$$

The term $\int_0^{X_R} \frac{\exp(+\alpha X^2)dX}{(1-X)^{2/3}}$ was evaluated from Equation 6 for tests in which no nitrogen pretreatment was used. For tests in which the feed coal char was initially devolatilized in nitrogen at the same temperature and pressure to be subsequently used for gasification in a gasifying atmosphere, reactivity for rapid-rate methane formation was destroyed above 1500°F and $X_R = 0$. For this condition

$$M(X) = \int_0^X \frac{\exp(+\alpha X^2)dX}{(1-X)^{2/3]}} = f_L k_T t \quad (22)$$

Typical plots of $M(X)$ vs. t are given in Figures 8–10 for data obtained with air-pretreated Ireland mine coal char. The slopes of the lines drawn correspond to values of $f_L k_T$ characteristic of each test.

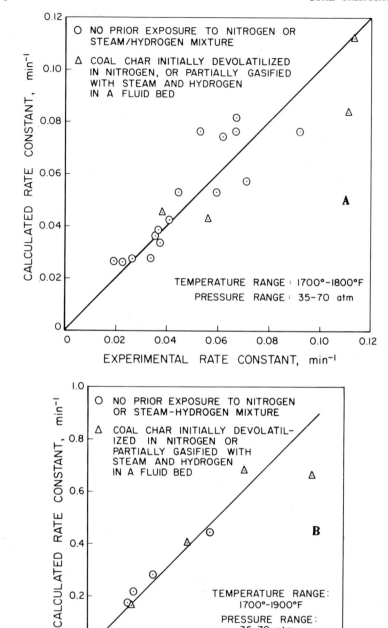

Figure 13. Comparison of experimental and calculated rate constants for low-rate gasification in steam–hydrogen mixture (thermobalance data)

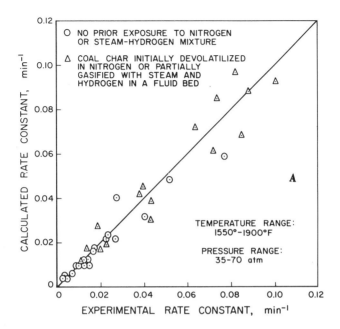
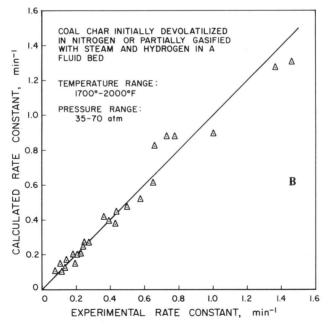

Figure 14. Comparison of experimental and calculated rate constants for low-rate gasification in synthesis gases (thermobalance data)

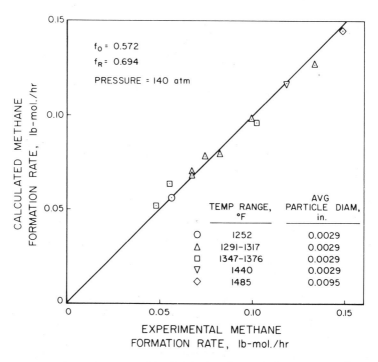

Figure 15. Experimental and calculated integral rates of methane formation for hydrogenation in a moving-bed reactor (IGT study)

Results of tests conducted with pure steam (Figure 8) were correlated using $\alpha = 0$ and were consistent with Equation 20 which corresponds to the situation in which specific gasification rates, $\frac{dX/dt}{(1-X)}$, increase with an increasing carbon conversion level. For gasification in hydrogen or steam–hydrogen mixtures (Figures 9 and 10), however, specific gasification rates generally decrease with increasing conversion level.

Carbon gasification rates were directly measured in the fluid-bed tests conducted with Disco char (1, 2, 3) and values of $f_L k_T$ and α can be obtained graphically by plotting values of $\ln \frac{dX/dt}{(1-X)^{2/3}}$ vs. X^2. Such a plot is illustrated in Figure 11, where $\frac{dX/dt}{(1-X)^{2/3}}$ at $X = 0$ is equal to $f_L k_T$, and the slope of a given line is equal to α. Generally the correlations presented are consistent with values of $f_L k_T$ and α obtained for the gasification of Disco char at 1600° and 1700°F and 1, 6, and 30 atm for gasification with pure hydrogen and with steam–hydrogen mixtures. The correlations are also consistent with individual rates of methane and carbon oxide formation reported in the studies done with Disco char for initial levels of carbon conversion. Although the relative rates of methane-to-carbon oxide formation were reported to increase somewhat with increasing conversion level (a trend not accounted for in the model developed in this study), investigators at the Consolidation Coal Co. have suggested that this effect was caused by a catalytic reaction downstream of the fluid-bed reactor used, in which some of the carbon monoxide produced in the reactor was converted to methane (17).

The consistency between the calculated and experimental values of $f_L k_T$ for tests conducted under various conditions with the thermobalance using air-pretreated Ireland mine coal char is demonstrated in Tables II-IV, and in Figures 12–14. The correlations developed have also been used to predict behavior in pilot-scale, moving- and fluid-bed tests conducted at IGT and elsewhere. The assumptions made in characterizing the nature of gas–solids contacting in these integral systems have been previously described (13). The most important assumptions made for fluid-bed systems are (a) the gas in the fluid bed is perfectly mixed, and (b) when continuous solids flow is used, the solids are in plug flow. For moving-bed systems we assumed that both gas and solids were in plug flow. With these simplifying assumptions, the conditions of primary

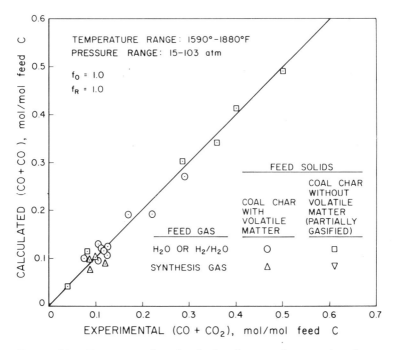

Figure 16. Experimental and calculated integral rates of carbon oxides formation for gasification in 2-, 4-, and 6-inch id fluid-bed reactors (IGT studies)

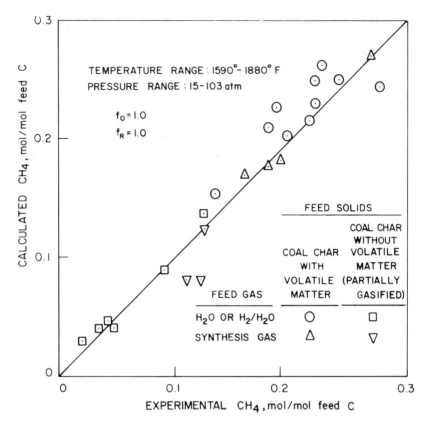

Figure 17. — Experimental and calculated integral rates of methane formation for gasification in 2-, 4-, and 6-inch id fluid-bed reactors (IGT studies)

importance in characterizing integral gasification behavior in individual tests include coal char feed rate and composition, particle residence times in the reactor, reactor temperature, pressure, feed gas composition, and flow rate. When coal char containing volatile matter was used as a feed material, rapid-rate methane formation and devolatilization were assumed to occur in a free-fall space above the reaction beds used. When devolatilized coal char was the feed material, no rapid-rate methane formation was considered to occur. Predicted and experimental integral rates of carbon oxides and methane formation are compared in Figures 15–19 which show good agreement for a wide range of experimental conditions.

Frequently, the $P_{CH_4}/P^2_{H_2}$ ratio in product gases from integral fluid-bed systems for the gasification of coal or coal char with steam-hydrogen-containing gases is greater than the equilibrium constant for the graphite-hydrogen–methane system. This has often been interpreted as corresponding to a situation in which the coal or coal char has a thermodynamic activity greater than unity with respect to graphite. The models proposed in this paper offer two other explanations for this phenomenon: Rapid-rate methane formation occurs when coal or coal char containing volatile matter is used as a feed material. The methane yield resulting from this step is kinetically determined, independent of methane partial pressure. Under certain conditions then, values of $P_{CH_4}/P^2_{H_2}$ greater than that corresponding to the equilibrium for the graphite–hydrogen–methane system can result. Values of $P_{CH_4}/P^2_{H_2}$ greater than that corresponding to the equilibrium considered can also occur for low-rate gasification of coal char according to the model assumed in this study. This is illustrated in Figure 20 where the results shown were based on computations of gas yields in a hypothetical fluid-bed for char gasification with a pure steam feed gas using the correlations described above. The reason for the behavior illustrated is that at intermediate values of hydrogen partial pressure, the rate of Reaction III, which produces methane, is greater

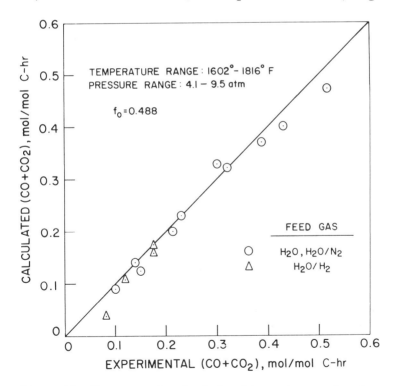

Figure 18. Experimental and calculated integral rates of carbon oxides formation for gasification in a fluid-bed reactor. Data from Ref. 18.

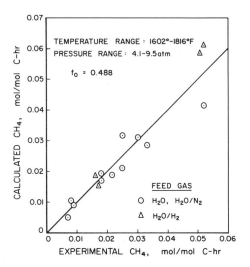

Figure 19. Experimental and calculated integral rates of methane formation for gasification in a fluid-bed reactor. Data from Ref. 18.

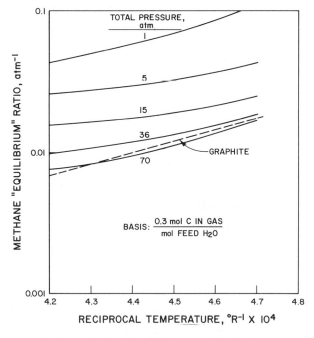

Figure 20. Calculated variations of the methane equilibrium ratio (P_{CH_2}/P_{H_2}) for gasification of carbon with steam in a backmixed reactor

than the reverse rate of Reaction II in which methane is consumed when a potential for carbon deposition by this reaction exists. The partial pressure dependencies defined in the correlations developed are such, however, that at sufficiently high hydrogen partial pressures Reaction II dominates, and equilibrium for this reaction is approached. The qualitative trends in Figure 20 and even the magnitudes of these trends bear a striking resemblance to a similar plot given by Squires (*19*) to correlate the activities of coals and chars for equilibrium in the char–hydrogen–methane system with temperature and pressure.

Definitions

A = nongasifiable matter in feed coal char (including ash and some sulfur), grams/gram feed coal char

C_b = base carbon in coal char at an intermediate gasification level, grams/gram feed coal char

$C_b°$ = carbon in the fixed carbon fraction of the feed coal char as determined by a proximate analysis, grams/gram feed coal char

$C_t°$ = total carbon in the feed coal char as determined by an ultimate analysis, grams/gram feed coal char

C_v = volatile carbon in feed coal, grams/gram feed coal char

f_o = relative reactivity factor for low-rate gasification dependent on the particular carbonaceous solid

f_L = relative reactivity factor for low-rate gasification which depends on the particular carbonaceous solid and pretreatment temperature

f_R = relative reactivity factor for rapid-rate methane formation dependent on the particular carbonaceous solid

k_T = overall rate constant for low-rate gasification, min^{-1}

$k_2(T), k_3(T)$ = rate constants, min^{-1}

k_I, k_{II}, k_{III} = rate constants for Reactions I, II, and III, min^{-1}

$K^E_I, K^E_{II}, K^E_{III}$ = equilibrium constants for Reactions I, II, and III, considering carbon as graphite

N_{A_o} = concentration of species A_o at any time, mole/mole base carbon

$N°_{A_o}$ = initial concentration of species A_o, mole/mole base carbon

N_{A*} = concentration of species A_* at any time, mole/mole base carbon

$P_{H_2}, P_{H_2O}, P_{CO}, P_{CO_2}, P_{CH_4}$ = partial pressures of H_2, H_2O, CO, CO_2, and CH_4, atm

t = time, min

T = reaction temperature, °R

T_o = pretreatment temperature, °R
V = volatile matter in feed coal char (including moisture), grams/gram feed coal char
W_o = weight of feed coal char, grams
ΔW = weight loss of coal char during gasification, grams
X = total base carbon conversion fraction
X_R = base carbon conversion fraction after reaction in rapid-rate methane formation stage is completed
α = kinetic parameter which depends on gas composition and pressure
β = $k_3(T)/k_2(T)$ ratio
γ = noncarbon matter evolved along with base carbon, grams/gram base carbon evolved

Literature Cited

1. Goring, G. E., et al., *Ind. Eng. Chem.* (1953) **45**, 2586.
2. Zielke, C. W., Gorin, E., *Ind. Eng. Chem.* (1955) **47**, 820.
3. Zielke, C. W., Gorin, E., *Ind. Eng. Chem.* (1957) **49**, 396.
4. Feldmann, H. F., Mima, J. A., Yavorsky, P. M., ADVAN. CHEM. SER. (1974) **131**, 108.
5. Zahradnik, R. I., Grace, R. J., ADVAN. CHEM. SER. (1974) **131**, 126.
6. Mosely, F., Patterson, D., *J. Inst. Fuel* (1965) **38**, 378.
7. Blackwood, J. D., McCarthy, D. J., *Aust. J. Chem.* (1966) **19**, 797.
8. Wen, C. Y., Huebler, J., *Ind. Eng. Chem., Process Des. Develop.* (1965) **4**, 147.
9. Eddinger, R. T., Freidman, L. D., Rau, A., *Fuel* (1966) **19**, 245.
10. Kimber, G. M., Gray, M. D., *J. Combust. Flame Inst.* (1967) **2**, 360.
11. Birch, T. J., Hail, K. R., Urie, R. W., *J. Inst. Fuel* (1960) **33**, 422.
12. Hiteshue, R. W., Friedman, S., Madden, R., *Bur. Mines Rep. Invest. No. 6376* (1964).
13. Pyrcioch, E. J., Feldkirchner, H. L., Tsaros, C. L., Johnson, J. L., Bair, W. G., Lee, B. S., Schora, F. C., Huebler, J., Linden, H. R., *IGT Res. Bull. No. 39*, Chicago, Nov. 1972.
14. Blackwood, J. D., *Aust. J. Chem.* (1959) **12**, 14.
15. Goring, G. E., et al., *Ind. Eng. Chem.* (1952) **44**, 1051.
16. Blackwood, J. D., McGrory, F., *Aust. J. Chem.* (1958) **11**, 16.
17. Curran, G., Gorin, E., *U.S. Off. Coal Res. R&D Rept. No. 16, Interim Rept. No. 3, Book 2*, Government Printing Office, Washington, D.C., 1970.
18. May, W. G., Mueller, R. H., Sweetser, S. B., *Ind. Eng. Chem.* (1958) **50**, 1289.
19. Squires, A. M., *Trans. Inst. Chem. Eng.* (1961) **39**, 3, 10, 16.

RECEIVED May 25, 1973. Work sponsored by the American Gas Association, U. S. Department of the Interior, Office of Coal Research and the Fuel Gas Associates.

11

Catalysis of Coal Gasification at Elevated Pressure

W. P. HAYNES, S. J. GASIOR, and A. J. FORNEY

Bureau of Mines, U. S. Department of the Interior, Pittsburgh, Pa. 15213

> *Various additives were evaluated for their catalytic effect on coal gasification. Steam–coal gasification tests were done in bench-scale units at 850°C and 300 psig with coal containing 5 wt % additive. Alkali metal compounds increased carbon gasification the most, by 31 to 66%. Twenty different metal oxides increased carbon gasification by 20–30%. Inserts coated with Raney nickel were active but lost activity rapidly. Pilot-plant tests were conducted in a Synthane gasifier at 907°–945°C and 40 atm. A 5% addition to the coal of either dolomite or hydrated lime resulted in significant increases in the amount of carbon gasified and in the amount of CH_4, CO, and H_2 produced.*

The gasification of coal with steam and oxygen under elevated pressure is an essential step in the Bureau of Mines Synthane process for converting coal to synthetic pipeline gas. A suitable catalyst or additive in the gasification step could conceivably improve the gasification of coal. Earlier investigators have catalyzed the gasification of coke and carbon with various additives and have demonstrated that some benefits would result from the catalysis of gasification at atmospheric pressure. Vignon (1), for example, showed that the per cent of methane in water gas made from coke increases significantly if lime is added to the coke. Neumann *et al.* (2) demonstrated the feasibility of improving the gasification of graphite at 450° to 1000°C by adding K_2O, CuO, and other salts. Continuation of the studies by Kroger and Melhorn (3) indicated that addition of either 8% Li_2O or (8% K_2O + 3% Co_3O_4) to low-temperature coke increased steam decomposition at 500° to 700°C. More recently, Kislykh and Shishakov (4) studied the effect of $Fe(CO)_5$, (Fe_2O_3 + $CuCl_2$), K_2CO_3, and NaCl on the fluid-bed gasification of wood charcoal at 750°C and atmospheric pressure. They found that sodium chloride was the most effective additive, accelerating gasification by 62% and increasing steam decomposition 2.5-fold.

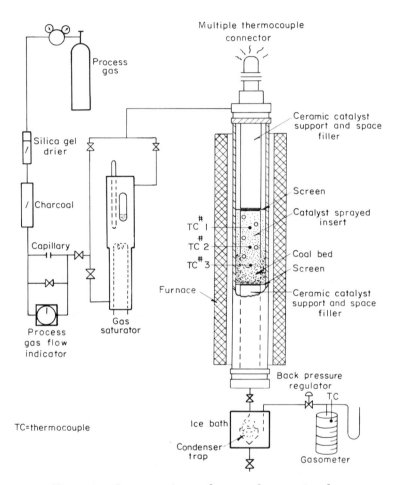

Figure 1. Apparatus for catalytic gasification of coal

The work on catalysis reviewed thus far does not include either gasification at elevated pressures or the use of volatile coals. The bench-scale work now reported compares the catalytic activity of various additives in the gasification of volatile coals with steam under pressure and studies other process parameters of interest such as gasification temperature, type of contact with the catalyst, degree of gasification, and repeated use of catalyst. Results of pilot plant gasification tests using additives are also reported.

Bench-Scale Studies

The gasification tests were conducted in two bench-scale reactor units, units A and B. The units were essentially the same; each unit contained an electrically heated reactor constructed of a 21-inch long section of 3/4-inch schedule 160 pipe of type 321 stainless steel. The coal-additive charge was contained in the middle 6 inches of the reactor to minimize the spread of bed temperatures. Alumina cylinders filled the void at each end of the reactor. Three thermocouples were located in the 6-inch reactor zone, 1/2 inch, 3 inches, and 5 1/2 inches from the top of the zone. The maximum temperature occurred at the middle thermocouple and was considered the nominal temperature of the reaction. The top and bottom bed temperatures were generally within 35°C of the maximum temperature. During a gasification test the coal charge was gasified by steam which was introduced by saturation of the nitrogen carrier gas.

Figure 1 shows a flow sheet of a reactor unit and its auxiliary equipment: high-pressure gas supply, silica gel and charcoal purifiers, calibrated capillary meter as feed gas flow indicator, gas saturator to humidify the feed gas, condenser and trap for liquid product collection, and gasometer for metering and collecting total product gas. Reactor pressure was controlled by a back-pressure regulator. Temperature of the gas saturation was controlled within ±0.6°C by a chromel–alumel thermocouple control system.

Analyses of dry product gases were done by mass spectrometer and gas chromatography. The liquid product, about 95% water, was drained and weighed.

Coal Used. The coal charged in all the tests discussed here was a single batch of high-volatile bituminous coal (Bruceton, Pa.) that had been pretreated at 450°C with a steam–air mixture to destroy its caking quality. The pretreated coal was crushed and sieved to a particle size of 20 to 60 mesh. Proximate and ultimate analyses of the pretreated coal are shown in Table I, in weight percent.

Table I. Analyses of Pretreated Coal Used for Feedstock

Proximate		Ultimate	
Moisture	1.5	Hydrogen	3.9
Volatile	22.4	Carbon	74.3
Fixed carbon	65.5	Nitrogen	1.5
Ash	10.6	Oxygen	8.7
		Sulfur	1.0
		Ash	10.6

Standard Gasification Tests for Screening Additives. Standard gasification tests were conducted in units A and B to determine the effect of various additives on the rate of carbon gasification, on the rate of gas production, and on other gasification parameters. In unit A, the standard gasification tests using various catalysts were made at the selected operating conditions of 850°C, 300 psig, and 5.8 grams/hr steam feed carried by 2000 cm^3 N_2/hr. The coal charge was 10 grams plus 0.5 gram of catalyst. All catalysts were either powders or crystals that were admixed

Table IIa. Specific Rate of Gas Production

Catalyst	Experiment No.	No. of Tests	Total Dry Gas Production Rate, N_2-Free Basis, std cc/hr/gram coal charged
No catalyst	166	6	331
Raney nickel, unactivated			
sprayed	203	6	453
mixed	197	3	414
Raney nickel, activated			
mixed	200	3	442
NaCl	209	3	471
KCl	212	3	615
ZnO	215	3	440
$NiCl_2.6H_2O$	216	3	442
$NiSO_4.6H_2O$	217	3	420
K_2CO_3	218	3	578
$ZnBr_2$	219	3	418
SnO_2	224	3	395
Fe	225	3	367

[a] Test conditions, unit A: charge, 10 grams pretreated Bruceton coal, 0.5 gram catalyst;

Table IIb. Specific Rate of Liquid Production, Production Using

Catalyst	Experiment No.	Product Liquid, gram/hr/gram coal charged	Carbon Gasification Rates, gram/hr/gram coal charged
No catalyst	167	0.478	89×10^{-3}
Raney nickel, unactivated			
sprayed	203	.472	98×10^{-3}
mixed	197	.501	106×10^{-3}
Raney nickel, activated			
mixed	200	.489	107×10^{-3}
NaCl	209	.506	117×10^{-3}
KCl	212	.480	148×10^{-3}
ZnO	215	.490	112×10^{-3}
$NiCl_2.6H_2O$	216	.493	110×10^{-3}
$NiSO_4.6H_2O$	217	.498	108×10^{-3}
K_2CO_3	218	.494	144×10^{-3}
$ZnBr_2$	219	.454	104×10^{-3}
SnO_2	224	.507	98×10^{-3}
Fe	225	.517	94×10^{-3}

[a] Test conditions, unit A: charge, 10 grams pretreated Bruceton coal, 0.5 gram catalyst;

Using Various Catalysts (Series A) [a]

Specific Production Rates of Constituent Gases, std cc/hr/gram coal charged

H_2	CO	CO_2	CH_4	C_2H_4	C_2H_6	C_3H_8	C_3H_6	Unit Heating Value N_2-Free Product Gas, Btu/scf	Gaseous Heating Value Produced, Btu/hr/ft³ coal charged
166	49.7	70.2	43.7	0.06	0.93	0.24	0.12	353	57,700
271	53.7	72.3	54.2	.12	.80	.22	.05	359	80,254
216	52.1	95.5	49.1	—	1.19	.18	—	336	68,712
243	47.9	104.5	45.3	.10	0.84	.13	—	322	70,258
254	53.0	115.5	47.6	.33	.55			318	74,032
340	90.1	137.3	46.7	.07	.64	.33	.03	307	93,250
232	52.2	104.4	50.0	.05	.92	.13	.13	330	71,718
237	51.7	103.6	48.4	.14	.82	.18	.16	328	71,716
220	48.0	102.9	48.0	.21	.78	.23	.05	328	68,163
309	95.0	126.4	46.2	.31	.72	.22	.06	312	89,011
224	53.9	96.2	42.6	.18	.79	.10	.08	324	66,783
213	37.7	98.1	44.0	.15	.65	.33	.28	326	63,700
193	39.1	88.3	45.4	.09	.95	.38	.24	339	61,500

feed, 5.8 grams H_2O/hr + 2000 std cc N_2/hr; approximately 4 hrs duration; 300 psig; 850°C.

Carbon Gasification, and Gaseous Heating Value Various Catalysts (Series A) [a]

feed, 5.8 grams H_2O/hr + 2000 std cc N_2/hr; approximately 4 hrs duration; 300 psig; 850°C.

Table IIIa. Specific Rate of Gas Production

Catalyst	Experiment No.	No. of Tests	Total Dry Gas Production Rate, N_2-Free Basis, std cc/hr/gram coal charged
No catalyst	300	3	297
$Ca(OH)_2$	301	3	345
Ni_2O_3	302	3	359
NiO	303	3	373
Ni	304	3	366
Fe_3O_4	305	3	380
CuO	306	3	385
MnO_2	307	3	375
BaO	308	3	395
ZrO_2	309	3	377
SrO	310	3	392
Bi_2O_3	311	2	390
Sb_2O_5	312	2	391
MgO	313	3	384
PbO_2	314	3	400
MoO_3	315	3	380
TiO_2	316	3	380
CrO_3	317	3	378
$LiCO_3$	318	3	439
V_2O_5	319	3	367
Cr_2O_3	320	3	374
Pb_3O_4	321	3	400
B_2O_3	322	3	394
Al_2O_3	323	3	380
CoO	324	4	383
Cu_2O	325	6	385

a Test conditions, unit B: charge, 10 grams pretreated Bruceton coal, 0.5 gram catalyst; 850°C.

with the coal except for experiment 203. In experiment 203 a metal tubular insert was flame-spray coated with about 10 grams of unactivated Raney nickel, then inserted in the coal bed. In experiment 200 the Raney nickel catalyst powder was activated or approximately 65% reduced by treatment with 2% sodium hydroxide solution. In the reduced activated state, the catalyst was dried, mixed with coal, and charged into the unit under an inert atmosphere of nitrogen.

Reaction time, at the desired reaction temperature, was held constant at 4 hrs. An additional heat-up time of about 40 min was needed to reach the desired reaction temperature. Steam flow was not started until bed temperature exceeded 200°C. Conditions in unit B were the same as in unit A except that the steam rate was slightly lower at 5.0 grams/hr. Generally, tests were conducted in triplicate or higher replication. The average deviations in carbon gasification rate determinations generally ranged from 1 to 5%.

Using Various Catalysts (Series B)[a]

Specific Production Rate of Constituent Gases,
std cc/hr/gram coal charged

H_2	CO	CO_2	CH_4	C_2H_4	C_2H_6	C_3H_8	C_3H_6
147	37.7	69.9	40.8	0.18	0.72	0.20	0.24
180	32.0	86.2	45.4	.09	.63	.12	.19
189	43.9	80.9	42.8	.07	.74	.73	.50
194	52.0	82.0	43.9	—	.94	.22	.29
189	53.6	76.6	46.2	.17	.46	.29	.27
195	60.3	78.0	45.6	.15	.49	.27	.22
201	56.3	79.9	46.8	.12	.52	.17	.07
193	45.7	90.2	44.9	.10	.79	.10	.15
202	48.9	97.1	45.5	.13	.73	.15	.13
192	57.8	80.0	46.2	.12	.88	.05	.12
203	48.2	91.3	47.6	.15	.76	.15	.15
200	63.9	80.4	44.4	.19	.72	.15	.04
201	63.6	79.7	45.9	.19	.45	.15	.15
199	48.4	87.7	47.7	.18	.72	.30	.23
210	52.7	89.1	47.7	.08	1.00	.18	.23
202	45.0	86.6	45.5	.03	1.09	.30	.13
197	48.9	85.7	47.2	.33	0.68	.13	.23
194	52.5	84.4	46.1	.07	.79	.07	.15
228	64.9	95.9	49.2	.19	.57	.27	—
188	47.7	83.5	45.9	.12	.71	.20	.12
184	58.6	83.0	47.4	.57	.57	.10	.12
205	57.2	87.3	49.1	.13	.72	.15	.13
203	61.0	79.0	47.4	.05	1.59	1.32	.08
196	54.6	82.2	46.7	.15	0.61	0.13	.09
201	48.6	87.0	45.4	.25	.43	.17	.07
203	53.1	84.4	44.1	.28	.42	.14	.13

feed, 5.0 grams H_2O/hr plus 2000 std cc N_2/hr; approximately 4 hrs duration; 300 psig;

The experimental results of the screening tests conducted in unit A are presented in Tables IIa and IIb, and for unit B they are given in Tables IIIa and IIIb. The specific gas production rates and gasification rates are based on the approximate 4-hr reaction time at 850°C. The results indicate that methane production rates as well as carbon gasification rates can be increased significantly if certain compounds are admixed with the coal feed.

The rate of carbon gasification for the uncatalyzed coal was about 10% higher in unit A than in unit B (experiment 167 vs. 300). The slightly higher steam rate, with its correspondingly higher partial pressure of steam and higher gas phase diffusion rate, is suspected as the cause of the higher reaction rate in unit A. Because of this difference in absolute rates, the percentage increases achieved in gasification rates and gas pro-

Table IIIb. Specific Rate of Liquid Production, Production Using

Catalyst	Experiment No.	Product Liquid, gram/hr/ gram coal charged	Carbon Gasification Rates, gram/hr/ gram coal charged
No catalyst	300	0.464	81×10^{-3}
$Ca(OH)_2$	301	.447	88×10^{-3}
Ni_2O_3	302	.439	92×10^{-3}
NiO	303	.422	97×10^{-3}
Ni	304	.413	96×10^{-3}
Fe_3O_4	305	.396	100×10^{-3}
CuO	306	.418	99×10^{-3}
MnO_2	307	.406	98×10^{-3}
BaO	308	.417	104×10^{-3}
ZrO_2	309	.405	99×10^{-3}
SrO	310	.417	101×10^{-3}
Bi_2O_3	311	.415	102×10^{-3}
Sb_2O_5	312	.414	102×10^{-3}
MgO	313	.391	100×10^{-3}
PbO_2	314	.397	103×10^{-3}
MoO_3	315	.438	96×10^{-3}
TiO_2	316	.364	99×10^{-3}
CrO_3	317	.398	99×10^{-3}
$LiCO_3$	318	.364	113×10^{-3}
V_2O_5	319	.402	96×10^{-3}
Cr_2O_3	320	.394	102×10^{-3}
Pb_3O_4	321	.410	105×10^{-3}
B_2O_3	322	.399	104×10^{-3}
Al_2O_3	323	.424	99×10^{-3}
CoO	324	.355	98×10^{-3}
Cu_2O	325	.363	98×10^{-3}

[a] Test conditions, unit B: charge, 10 grams pretreated Bruceton coal, 0.5 gram catalyst; 850°C.

duction rates resulting from additives were related only to rates obtained in the gasification of the uncatalyzed coal in the same unit. Shown in Table IV are the relative effects of 40 additives on the production of methane and hydrogen; Table V shows their relative effects on the production of carbon monoxide and the gasification of carbon. The additives are listed in decreasing order of per cent increased production. The corresponding reactor unit used (either A or B) is also listed.

Table IV shows that at standard test conditions all the additives listed except $ZnBr_2$ increased methane production and that all the tested additives increased hydrogen production significantly. The sprayed Raney nickel catalyst increased methane production by 24% and was the most effective material for promoting methane production. The next three

Carbon Gasification, and Gaseous Heating Value
Various Catalysts (Series B)[a]

Unit Heating Value N_2-Free Product Gas, Btu/scf	Gaseous Heating Value Produced, $Btu/hr/ft^3$ coal charged
350	51,300
339	57,700
344	60,960
340	62,700
349	63,100
345	64,800
344	65,500
333	61,700
328	64,050
344	64,200
337	65,200
340	65,500
343	66,200
342	64,900
339	67,000
339	63,800
342	64,200
340	63,400
334	72,400
341	61,900
345	63,700
342	67,600
355	69,200
342	64,300
336	63,800
336	63,900

feed, 5.0 grams H_2O/hr plus 2000 std cc N_2/hr; approximately 4 hrs duration; 300 psig;

materials ranking highest in promoting methane production were $LiCO_3$, Pb_3O_4, and Fe_3O_4 with respective methane increases of 21, 20, and 18%. A comparison of the methanation activity of zinc oxide with that of zinc bromide indicates that the anion group of a catalyst can exert significant influence on the activity.

As shown in Table IV the alkali metal compounds were among the best promoters of hydrogen production. The per cent increases in hydrogen produced were 105, 83, 55, and 54%, respectively, with the addition of KCl, K_2O_3, $LiCO_3$, and NaCl. The sprayed Raney nickel was the third most effective promoter of hydrogen production, yielding a hydrogen increase of 63%. Table V shows that the alkali metal compounds K_2CO_3, KCl, and $LiCO_3$ gave the greatest increase in carbon monoxide production as well as in carbon gasification; the increase ranged from 40 to 91%.

Table IV. Increase in the Production of Methane and Hydrogen

Catalyst	Unit	Increase in CH_4, %	Catalyst	Unit	Increase in H_2 Produced, %
Raney nickel unactivated spray	A	24	KCl	A	105
$LiCO_3$	B	21	K_2CO_3	A	83
Pb_3O_4	B	20	Raney nickel, unactivated spray	A	63
Fe_3O_4	B	18	$LiCO_3$	B	55
MgO	B	17	NaCl	A	53
PbO_2	B	17	Raney nickel, activated mix	A	46
SrO	B	17	$NiCl_2 \cdot 6H_2O$	A	43
TiO_2	B	16	PbO_2	B	43
Cr_2O_3	B	16	ZnO	A	40
B_2O_3	B	16	Pb_3O_4	B	39
CuO	B	15	SrO	B	38
ZnO	A	14	B_2O_3	B	38
Al_2O_3	B	14	Cu_2O	B	38
Ni	B	13	CuO	B	37
ZrO_2	B	13	BaO	B	37
Sb_2O_5	B	13	Sb_2O_5	B	37
CrO_3	B	13	MoO_3	B	37
V_2O_5	B	13	CoO	B	37
Raney nickel, unactivated mix	A	12	Bi_2O_3	B	36
BaO	B	12	$ZnBr_2$	A	35
MoO_3	B	12	MgO	B	35
$NiCl_2 \cdot 6H_2O$	A	11	TiO_2	B	34
$Ca(OH)_2$	B	11	$NiSO_4 \cdot 6H_2O$	A	33
CoO	B	11	Fe_3O_4	B	33
$NiSO_4 \cdot 6H_2O$	A	10	Al_2O_3	B	33
MnO_2	B	10	NiO	B	32
NaCl	A	9	CrO_3	B	32
Bi_2O_3	B	9	MnO_2	B	31
NiO	B	8	ZrO_2	B	31
Cu_2O	B	8	Raney nickel, unactivated mix	A	30
KCl	A	7	Ni_2O_3	B	29
K_2CO_3	A	6	Ni	B	29
Ni_2O_3	B	5	SnO_2	A	28
Raney nickel, activated mix	A	4	V_2O_5	B	28
Fe	A	4	Cr_2O_3	B	25
SnO_2	A	1	$Ca(OH)_2$	B	22
$ZnBr_2$	A	−3	Fe	A	16

Table V. Increase in Production of Carbon Monoxide and Gasification of Carbon

Catalyst	Unit	Increase in CO produced, %	Catalyst	Unit	Increase in Carbon Gasified, %
K_2CO_3	A	91	KCl	A	66
KCl	A	81	K_2CO_3	A	62
$LiCO_3$	B	72	$LiCO_3$	B	40
Bi_2O_3	B	69	NaCl	A	31
Sb_2O_5	B	69	Pb_3O_4	B	30
B_2O_3	B	62	BaO	B	28
Fe_3O_4	B	60	B_2O_3	B	28
Cr_2O_3	B	55	PbO_2	B	27
ZrO_2	B	53	Bi_2O_3	B	26
Pb_3O_4	B	52	Cr_2O_3	B	26
CuO	B	49	Sb_2O_5	B	26
Al_2O_3	B	45	ZnO	A	26
Ni	B	42	SrO	B	25
Cu_2O	B	41	$NiCl_2 \cdot 6H_2O$	A	24
PbO_2	B	40	MgO	B	23
CrO_3	B	39	Fe_3O_4	B	23
NiO	B	38	CuO	B	22
BaO	B	30	ZrO_2	B	22
TiO_2	B	30	TiO_2	B	22
CoO	B	29	CrO_3	B	22
MgO	B	28	Al_2O_3	B	22
V_2O_5	B	27	$NiSO_4 \cdot 6H_2O$	A	21
MnO_2	B	21	MnO_2	B	21
MoO_3	B	19	CoO	B	21
Ni_2O_3	B	16	Cu_2O	B	21
SrO	B	15	Raney nickel, activated mix	A	20
Raney nickel, unactivated spray	A	8	NiO	B	20
$ZnBr_2$	A	8	Raney nickel, unactivated mix	A	19
NaCl	A	7	Ni	B	19
Raney nickel, unactivated mix	A	5	MoO_3	B	19
ZnO	A	5	V_2O_5	B	19
$NiCl_2 \cdot 6H_2O$	A	4	$ZnBr_2$	A	17
$NiSO_4 \cdot 6H_2O$	A	−3	Ni_2O_3	B	14
Raney nickel, activated mix	A	−4	Raney nickel, unactivated spray	A	10
$Ca(OH)_2$	B	−15	SnO_2	A	10
Fe	A	−21	$Ca(OH)_2$	B	9
SnO_2	A	−24	Fe	A	6

NaCl gave a significant increase of 31% in carbon gasification but was much less effective in increasing the production of carbon monoxide (7% increase).

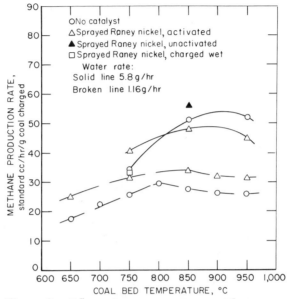

Figure 2. Effect of temperature on methane production rate using sprayed Raney nickel. Test conditions: pressure, 300 psig; flow, 2000 std cc/hr N_2; water, 5.8 and 1.16 grams/hr.

As shown in Tables IIb and IIIb the unit heating values of the total product gases generally decreased if additives were mixed with the coal, but since the total gas make was increased, the total amount of fuel value produced as product gas increased. Two methods of adding catalyst—by admixing with the coal or by coating the surface of an insert or carrier—may be compared for unactivated Ranel nickel catalyst (experiments 203 and 197). Admixing the catalyst with the coal provides better contact between coal and catalyst than does the insertion of catalyzed surfaces into the bed of coal. The superior contact achieved by admixing the catalyst and coal is proved by the higher specific rate of carbon gasification obtained by the admixed Raney nickel (unactivated) in experiment 197 over that of sprayed Raney nickel (unactivated) in experiment 203 (Tables IIa and IIb). This is further confirmed analytically by the larger amount of carbon left in the residue with the sprayed Raney nickel catalyst.

Table IIa indicates that the Raney nickel (unactivated) catalyst was more effective in promoting methane production when sprayed on a surface than when it was admixed with the coal charge. Apparently the process of methane synthesis from CO and H_2 was more effectively promoted by the sprayed Rany nickel (unactivated) catalyst. Another possible reason for the greater production of methane with the sprayed Raney nickel (unactivated) catalyst is that the poorer contact between coal and catalyst resulted in less reforming of methane and other hydrocarbons. Hydrogen production was also greater when the sprayed Raney nickel (unactivated) catalyst was used than when Raney nickel (unactivated) was admixed with coal.

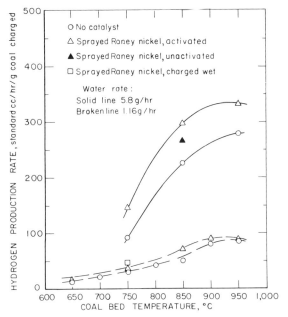

Figure 3. Effect of temperature on specific production of hydrogen, using sprayed Raney nickel. Test conditions: pressure, 300 psig; flow, 2000 std cc/hr N_2; water, 5.8 and 1.16 grams/hr.

Effect of Temperature and Steam Rate. Gasification experiments similar to the standard tests were conducted in unit A except that the reaction temperatures were varied from 650° to 950°C and steam rates used were 1.16 and 5.8 grams/hr. The catalyst used was flame-sprayed Raney nickel 65% activated. Also, tests were conducted at 750°C and 1.16 grams/hr steam rate with sprayed Ranel nickel activated and charged wet and at 850°C and 5.8 grams/hr steam rate with sprayed Raney nickel unactivated.

Results of these experiments are in Figure 2, 3, 4, 5, and 6 where production rates of methane, hydrogen, carbon monoxide, carbon gasifica-

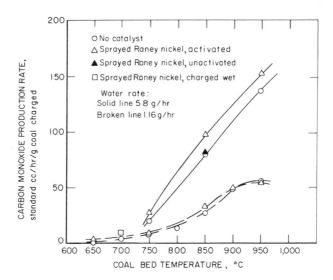

Figure 4. Effect of temperature on carbon monoxide production rate using sprayed Raney nickel. Test conditions: pressure, 300 psig; flow, 2000 std cc/hr N_2; water, 5.8 and 1.16 grams/hr.

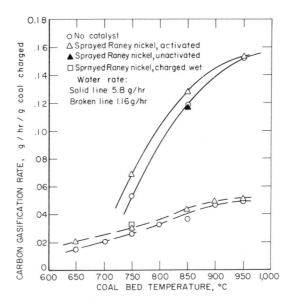

Figure 5. Effect of temperature on carbon gasification rate using sprayed Raney nickel. Test conditions: pressure, 300 psig; flow, 2000 std cc/hr N_2; water, 5.8 and 1.16 grams/hr.

tion, and production of total dry gas, respectively, are shown. At the temperatures and steam rates shown in these figures the presence of the sprayed Raney nickel insert has resulted in a higher production of nearly all the major gases and in a higher carbon gasification than that achieved without catalysts. One exception is methane production with the activated sprayed Raney nickel at 850° and 950°C for 5.8 grams/hr steam rate (Figure 2) when methane production was greater for the uncatalyzed reaction than for the catalyzed reaction. However, a 6% increase in methane production was achieved at 850°C and 5.8 grams/hr steam rate when the unactivated sprayed Raney nickel was used as compared with the experiment when no catalyst was used (Figure 2). The other exception is the production of carbon monoxide at 950°C and 1.16 grams/hr steam rate (Figure 4). In this case, carbon monoxide production was 5% lower for the catalyzed reaction than for the uncatalyzed reaction.

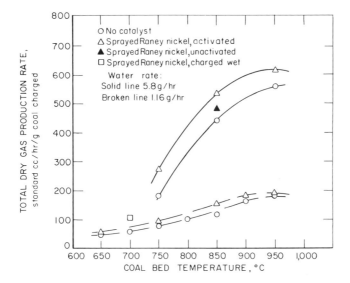

Figure 6. Effect of temperature on total dry gas production rate using sprayed Raney nickel. Test conditions: pressure, 300 psig; flow, 2000 std cc/hr N_2; water, 5.8 and 1.16 grams/hr.

In general the effectiveness of the catalyst decreases as temperature is increased. For example, the increases achieved in carbon gasification and in total gas production attributable to the catalyst became smaller as the reaction temperature increased from 750° to 950°C (Figures 5 and 6). In the 5.8 grams/hr steam rate tests, the carbon gasification rate at 750°C was increased by 0.017 gram/hr/gram coal charged for a 33% increase whereas at 950°C the increase in carbon gasification was negligible at an

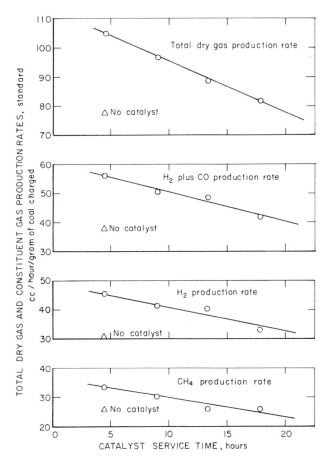

Figure 7. Effect of catalyst service time on gas production rates. Test conditions: temperature, 750°C; flow, 2000 std cc/hr N_2 and 1.16 grams/hr water.

increment of 0.001 gram/hr/gram coal charged. Similarly, total gas production was increased 43% at 750°C and 10% at 950°C. As shown by the gas production and carbon gasification rates in Figures 2–6, for reaction temperatures of 750°C and higher, the higher steam feed rate of 5.8 grams/hr resulted in significantly higher reaction rates and in larger increases in such rates caused by catalysis as compared with those achieved at the lower steam feed rate of 1.16 grams/hr. In the case of the lower steam feed rate, it is possible that much of the potential increase in reaction rate attributable to catalysis may have been masked because of a low diffusion rate in the gas phase.

Repeated Use of Sprayed Raney Nickel Catalyst. To determine the stability of sprayed Raney nickel catalyst, a single insert flame sprayed with Raney nickel catalyst was subjected sequentially to four gasification tests at 750°C, 300 psig, and a steam rate of 1.16 grams/hr. A fresh charge of coal was used in each run. Duration of each run was 4 to 5 hours. The resulting gas production rates are shown in Figure 7; also shown is the base case with no catalyst inserted in the bed.

Gas production rates in Figure 7 indicate that the activity of the sprayed Raney nickel catalyst insert decreased rapidly with use. The gas production of the fourth test (17.8-hr service) was only about 10% greater than that obtained when no catalyst was used. Extrapolation of the total gas production rate as a function of accumulated service time on one catalyst insert indicates that the catalyst insert would be completely ineffective after about 20 hours of operation. Considerable flaking-off of the catalyst is apparent; thus the need is indicated for an effective bonding agent or alloying substance that will increase the physical durability of the catalyst. Sulfur compounds gasified from the coal are also suspected of poisoning the nickel catalyst and resulting in the decline in activity with time.

Effect of Extent of Gasification Time. Tests were conducted to determine whether catalysts remain effective as the extent of gasification increases. Sprayed Raney nickel inserts and CaO powder were subjected to the standard test conditions of 850°C, 300 psig, 5.8 grams/hr steam rate but with gasification times varying from 2 to 8 hrs. The abridged results in Table VI show that although the overall rate of gasification decreases with extent of gasification or with gasification time, the catalysts tested still generally increased the rate of carbon gasification and the rate of gas production, as indicated by CH_4 production, over that achieved with no catalyst.

Effect of Residues from Catalytic Gasification. The catalytic effectiveness of ash residues (some contain either residual KCl or K_2CO_3) from total gasification operations were tested and compared with the effectiveness of the fresh additive, K_2CO_3 and KCl. The gasification residues were prepared by nominally complete steam-gasification of coal in a 1-inch diameter electrically heated, vertically mounted stainless steel reactor. The coal charge consisted of 70 grams of pretreated Bruceton coal plus 3.5 grams of admixed catalyst; the charge was gasified at 950° to 970°C and atmospheric pressure using a steam feed rate of 45 grams/hr. The preparatory gasification step was stopped whenever the flow of dry product gas appeared to cease. Carbon and ash content of the pretreated coal and of residues after total gasification and residue analyses are presented in Table VII. The ash analyses show that ash from the catalyzed coals contained significantly larger amounts of potassium than did the ash of uncatalyzed coal.

Table VI. Effect of Extent of Gasification

Gasification Time, hr	CH_4 Production	
	2	6
Rates with no catalyst	53.3 cc/hr/gram	28.6 cc/hr/gram
Rates with CaO	56.9 cc/hr/gram	30.7 cc/hr/gram
Increase caused by CaO	7%	7%
Rates with sprayed Raney nickel catalyst	70.9 cc/hr/gram	38.3 cc/hr/gram
Increase caused by Raney nickel	33%	34%

[a] Unit A: temperature, 850° C; N_2 flow, 2000 std cc/hr; steam rate, 5.8 grams/hr.

Table VII. Carbon, Ash Content, and Ash Analysis of

	Experiment No.	Carbon and Ash Content, %			
		Carbon	Ash	Al_2O_3	CaO
Charge coal	C-1	74.3	10.6	25.9	1.7
Residues					
no catalyst	D-1	1.67	98.9	25.1	1.6
K_2CO_3	D-2	0.47	99.9	19.7	1.1
KCl	D-3	4.7	95.1	22.2	1.3

[a] Gasification conditions: charge, 70 grams pretreated Bruceton coal, 3.5 grams of

Table VIII. Specific Production and Gasification Rates Resulting from

Experiment No.	No. of Tests	Residue Charged No.	Residue Charged grams	Total Dry Gas Production Rate, N_2-Free Basis, std cc/hr/gram Coal Charged
No catalyst	167	6		331
Catalyst-free residue	220	3	D-1	1.12	344
K_2CO_3	218	3			578
K_2CO_3 residue	221	3	D-2	1.41	398
KCl	212	3			615
KCl residue	222	3	D-3	1.26	370

	Experiment No.	Product Liquid, gram/hr/gram Coal Charged	Carbon Gasification Rates, gram/hr/gram Coal Charged
No catalyst	167	0.478	89×10^{-3}
Catalyst-free residue	220	.449	88×10^{-3}
K_2CO_3	218	.494	144×10^{-3}
K_2CO_3 residue	221	.487	104×10^{-3}
KCl	212	.480	148×10^{-3}
KCl residue	222	.469	104×10^{-3}

[a] Standard test conditions, unit A: 10 grams pretreated Bruceton coal, 0.5 gram catalyst 4 hrs duration; 300 psig; 850°C.

Time on Overall Effectiveness of Catalyst[a]

Carbon Gasification	
2	6
0.087 gram/hr/gram	0.062 gram/hr/gram
.090 gram/hr/gram	.062 gram/hr/gram
3%	0%
.093 gram/hr/gram	.063 gram/hr/gram
7%	2%

Pretreated Bruceton Coal and of Residues from Total Gasification[a]

Mineral Analysis of Ash, %							
Fe_2O_3	MgO	P_2O_5	K_2O	SiO_2	Na_2O	SO_3	TiO_2
10.8	0.9	<0.01	1.7	54.8	1.6	1.6	1.0
14.4	.8	—	1.9	54.7	1.5	—	—
10.8	.8	—	24.3	41.7	1.6	—	—
12.4	.7	—	13.1	48.5	1.4	0.4	—

catalyst; water rate, 45 cc/hr; 950°–970° C; atmospheric pressure.

Addition of Potassium Compounds and Various Gasification Residues[a]

Specific Production Rate of Constituent Gases, std cc/hr/gram coal charged							
H_2	CO	CO_2	CH_4	C_2H_4	C_2H_6	C_3H_8	C_3H_6
166	49.7	70.2	43.7	0.06	0.93	0.24	0.12
180	39.2	79.3	43.9	.07	.81	.33	.21
309	95.0	126.4	46.2	.31	.72	.22	.06
205	45.8	98.1	47.6	.27	.72	.12	.30
340	90.1	137.3	46.7	.07	.64	.33	.03
175	47.8	94.7	48.6	.22	.72	.19	.24

Unit Heating Value N_2-Free Product Gas, Btu/scf	Gaseous Heating Value Produced, Btu/hr/cu ft coal charged
353	57,700
344	58,500
312	89,000
332	65,300
307	93,300
336	61,300

(residue charge as indicated); feed 5.8 grams H_2O/hr + 2000 std cc N_2/hr; approximately

In the standard screening tests conducted in unit A the amount of residue admixed with the 10-gram coal charge was one-seventh of the residue from the total gasification. Thus, a theoretical equivalent of 0.5 gram of the original 3.5 grams of catalyst was charged with the coal in the standard screening test. Production rates obtained in the standard screening test are given in Table VIII for the case of plain pretreated coal and for the cases of addition of residues from the total gasification of the coals admixed with K_2CO_3 and KCl.

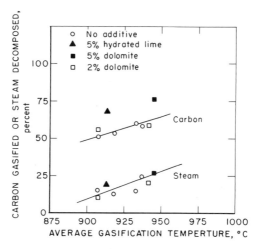

Figure 8. Effect of additives on per cent carbon gasified and per cent steam decomposition in 4-inch Synthane gasifier operating at 40 atm

The results of the tests on residues from total gasification of coal indicate that potassium compounds (K_2CO_3 and KCl) in the residues retained most of their activity in increasing the production of methane, lost part of their capability of increasing hydrogen production, and inhibited carbon monoxide production. The addition of 1.14 grams of catalyst-free residue had very little effect on either methane or total gas production (experiment 220 vs. 167).

Pilot-Plant Tests

Tests using additives mixed in the coal feed were conducted in the Bureau's 4-inch diameter Synthane gasifier system. In this system coal is first decaked in a fluidized-bed pretreater and then dropped into the fluidized-bed gasifier for steam–oxygen gasification. The general operation has been described by Forney et al. (5).

Ranges of pretreater and gasifier conditions used in this series of tests were as follows:

		Pretreater	Gasifier
Coal rate,	lbs/hr	17.8 – 21.2	— —
O_2 feed,	scf/lb coal	.31 – .37	2.12 – 3.4
N_2 feed,	scf/lb coal	5.4 – 6.2	— —
Steam feed, scf/lb coal		— —	19.6 – 25.4
Av. temp.,	°C	388 – 515	907 – 945

Pretreater and gasifier pressure was 40 atm.

To alleviate operating difficulties in the pretreater caused by unwanted steam condensation, a nitrogen gas feed was substituted for the steam feed of this pretreater. The coal feed was Illinois No. 6, River King Mine, 20 × 0 mesh. Additives were hydrated lime and dolomite, and additive concentrations used in the coal feed mixtures were 5 and 2 wt %, respectively. The analyses and sizes of the additives were as follows:

Hydrated lime: minimum CaO, 72 wt %
minimum MgO, .05 wt %
95% less than 325 mesh, and

Dolomite: $CaCO_3$, 55 wt %
$MgCO_3$, 44 wt %
85% less than 100 mesh

Results. The effect of the additives on the per cent carbon gasified and per cent steam decomposed can be seen in Figure 8 for gasification temperatures in the 900°–950°C range. Addition of 5% of either hydrated lime or dolomite to the coal resulted in significant increases in the amount of carbon gasified. The 5% hydrated lime addition resulted in an increase in the amount of carbon gasified of about 29% at an average gasification temperature of 914°C. These increases compare favorably with the 9% increase obtained in the 850°C bench-scale test using 5% addition of hydrated lime. The 2% addition of dolomite did not significantly increase the per cent carbon gasification. Steam decomposition was not significantly increased by either dolomite or hydrated lime at the 2 and 5% levels, respectively.

The effect of the additive upon yield of hydrogen and methane in the pilot-plant unit is shown in Figure 9. At an average gasification temperature of 914°C, addition of 5% hydrated lime in the coal feed increased the hydrogen yield approximately 30% from 6.25 to 8.1 scf/lb of moisture-and-ash-free coal feed. A similar increase of 17% was obtained when 5% dolomite was used in the coal feed at 945°C average gasification temperature. Such significant increases in hydrogen production were also observed in the bench-scale tests.

The yield of methane was increased significantly by 25% at 914°C average gasification temperature by adding 5% hydrated lime. The use of 5% dolomite at 945°C gasification temperature brought no significant increase in methane yield. Failure of catalytic action to increase the yield of methane at the higher gasification temperature (945°C) agrees with the general trend observed on bench-scale tests—i.e., that the increases in gas yield from catalysis decreases with an increase in temperature.

The effect of additives on the yield of carbon monoxide is shown in Figure 10. Addition of hydrated lime and dolomite at the 5% concentrations brought respective increases of 23 and 26% in yield of carbon monoxide. Figure 11 shows similar increase in product gas yield (CO + H_2 + CH_4) for the same additions. Adding 2% dolomite failed to bring any significant increase in yields of methane, hydrogen, or carbon monoxide.

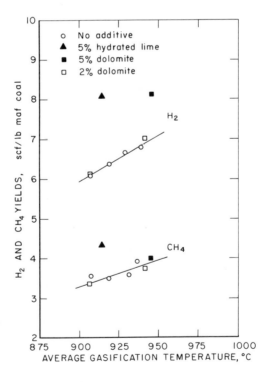

Figure 9. Effect of additives on yield of hydrogen and methane in 4-inch Synthane gasifier operating at 40 atm

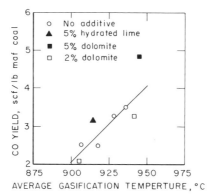

Figure 10. Effect of additives on yield of carbon monoxide in 4-inch Synthane gasifier operating at 40 atm

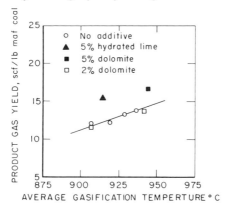

Figure 11. Effect of additives on yield of product gas $(CO + H_2 + CH_4)$ in 4-inch Synthane gasifier operating at 40 atm

Other results in pilot-plant operation caused by dolomite and hydrated lime at the 2 and 5% concentration levels, respectively, are that higher peak temperatures could be tolerated in the gasifier without incurring excessive sintering. With no additive in the coal, if a local temperature in the gasifier exceeded 1000°C, the operation generally would terminate because of excessive sintering or slagging of the char ash. With the additives in the coal, local temperatures as high as 1045°C were encountered with no adverse effect on operations. A similar elevation in sintering temperature induced by adding limestone to the feed coal is reported to be a key feature of the high-temperature Winkler process (6).

Summary and Conclusions

The overall results of the bench-scale investigation suggest that suitable additives would improve the coal gasification reaction at elevated pressures. In connection with possible benefits to the Synthane process for making high-Btu gas, it appears that appropriate additives could significantly increase production of methane and hydrogen in the gasification step. The bench-scale study thus far has shown the following:

(1) Alkali metal compounds and many other materials such as oxides of iron, calcium, magnesium, and zinc significantly increase the rate of carbon gasification and the production of desirable gases such as methane, hydrogen, and generally carbon monoxide during steam–coal gasification at 850°C and 300 psig.

(2) The greatest yield of methane occurred using an insert which was flame-sprayed with Raney nickel catalyst (unactivated). The Raney nickel had a limited activity life. Significant methane increase resulted from the addition of 5 wt % $LiCO_3$, Pb_3O_4, MgO, and many other materials.

(3) Increased gasification resulted whether the extent of coal gasification was small or great.

(4) At temperatures above 750°C catalytic effectiveness decreased with further increase in temperature.

(5) Residue from total gasification of coal mixed with potassium compounds still contained a significant concentration of potassium (over 10%) and was effective as an additive in increasing production of hydrogen and methane.

Operation of the 4-inch diameter Synthane pilot-plant gasifier at 40 atm pressure and average temperature of up to 945°C with dolomite and hydrated lime additives at 5% concentration has increased product gas ($CO + H_2 + CH_4$) yield significantly and has increased allowable operating temperatures.

Literature Cited

1. Vignon, L., *Ann. Chim.* (1921) **15**, 42–60.
2. Neumann, B., Kroger, C., Fingas, E., *Z. Anorg. Allgem. Chem.* (1931) **197**, 321–338.
3. Kroger, C., Melhorn, Gunter, *Brennstoff-Chem.* (1938) **19**, 257–262.
4. Kislykh, V. I., Shishakov, N. V., *Gaz. Prom.* (1960) **5**, 15–19.
5. Forney, A. J., Haynes, W. P., Corey, R. C., "The Present Status of the Synthane Coal-to-Gas Process," *Soc. Petrol. Engrs., Ann. Fall Meetg., 46th, Prepr.,* SPE 3586 (Oct. 3-6, 1971).
6. Meraikib, M., Franke, F. H., *Chem. Eng. Tech.* (1970) **42**, 834–836.

RECEIVED May 25, 1973.

12

Alkali Carbonate and Nickel Catalysis of Coal–Steam Gasification

W. G. WILLSON, L. J. SEALOCK, JR., F. C. HOODMAKER,
R. W. HOFFMAN, D. L. STINSON, and J. L. COX

Department of Mineral Engineering, University of Wyoming, Laramie, Wyo. 82071

The results of catalysis of coal–steam reactions with potassium carbonate, nickel, and a combination of the two catalysts in a single-stage batch charge reactor are compared with the thermal conversion. The gasification rate and coal conversion are promoted by the alkali carbonate while the nickel functions to methanate the carbon oxides and hydrocrack the liquids that are produced in its absence. About 60% carbon conversion is effected by this system at 650°C and 2 atm pressure. A gaseous product is produced with a CO_2-free heating value of 850 Btu per scf. Furthermore, it has been established that 20 wt % K_2CO_3 produces optimum gasification results. The form of the potassium in the ash is the same as initially charged.

Considerable work has been done on the catalytic effects of various chemicals on the reactions of carbon with steam. This work is of classic importance in the water–gas reactions. In addition, many studies have been made to determine active catalysts for synthesizing hydrocarbons from carbon monoxide and hydrogen. These two reaction systems are usually carried out at vastly different conditions of temperature and pressure.

The carbon–steam reactions represented by $C + H_2O = CO + H_2$ and $C + 2H_2O = CO_2 + 2H_2$ are endothermic, and even in the presence of catalysts the operating temperature range of 800°–1100°C is normally used (1). One of the earliest investigations of catalysts for the carbon–steam reactions carried out by Taylor and Neville (2) was done at 490°–570°C. Their most effective catalyst used with steam and coconut charcoal was potassium carbonate although sodium carbonate also proved effective. Fox and White (1) demonstrated the catalytic effect of impregnating graphite with sodium carbonate over the range 750°–1000°C.

Kröger (3) found that metallic oxides and alkali carbonates or mixtures catalyzed the carbon–steam reactions. Lewis and co-workers (4) stated that if reactive carbons are catalyzed with alkali carbonates, reasonable gasification rates are attainable at temperatures as low as 650°C. A process which uses molten sodium carbonate to catalyze as well as to supply heat for the carbon–steam gasification has been described(5).

In contrast to the carbon–steam reactions, hydrocarbons are usually synthesized from carbon monoxide and hydrogen below 450°C (6). The most active catalysts are group VIII metals mixed with various activating materials (7). A method for producing hydrocarbons directly from coal–steam systems using multiple catalysts in a single-stage reactor has been described by Hoffman (8). He and his co-workers (9) have described the effects of various commercial nickel methanation catalysts in a single-stage reactor. Nickel was chosen since the hydrocarbon yield was limited essentially to methane.

In the single-stage reactor the most effective mixed catalysts for producing methane and carbon dioxide from coal–steam systems ($2C + 2H_2O = CH_4 + CO_2$) are potassium carbonate and nickel. To get effective contact time in the single-stage reactor, the coal-to-nickel catalyst ratio must be about 1:1. With this in mind it was significant to determine the optimum ratio of potassium carbonate to coal which would give the best methane production.

The principal objectives of this investigation were (a) to demonstrate the feasibility of direct methane production by a single-stage multiple catalyst conversion, (b) to determine the optimum ratio of potassium carbonate to coal, holding the nickel catalyst concentration constant, (c) to determine by analytical methods and x-ray diffraction the form and amount of the potassium in the ash, and (d) to examine its water solubility.

Experimental

Feed Materials. The coal used in all runs was sub-bituminous from Glenrock, Wyo., ground to 60–100 mesh. Its analysis is given in Table I. ACS analytical reagent grade anhydrous potassium carbonate was used as the alkali catalyst. It was approximately the same mesh size as the coal. X-ray studies indicate, however, that some of the potassium carbonate had become hydrated. In addition, a commercial nickel catalyst was used. The nickel methanation catalyst (Ni-3210) containing approximately 35% by weight nickel on a proprietary support was purchased from the Harshaw Chemical Co. It was reduced with H_2 at *ca.* 650°C for 12–18 hours and stored under a nitrogen atmosphere.

Table I. Analysis of Glenrock Coal

	Proximate Analysis	
	As Received	Moisture Free
Moisture, wt %	12.2	—
Volatile matter, wt %	39.6	45.1
Fixed carbon, wt %	36.1	41.1
Ash, wt %	12.1	13.8
Heating value, Btu/lb	9140	10410

	Ultimate Analysis	
	As Received	Moisture Free
Hydrogen, wt %	5.1	4.3
Carbon, wt %	52.7	60.0
Nitrogen, wt %	0.6	0.7
Oxygen, wt %	28.6	20.2
Sulfur, wt %	0.8	1.0
Ash, wt %	12.1	13.8

Gas Analysis. Product gas volumes were measured by a calibrated wet test meter. Gas compositions were determined with a Beckman model GC-5 dual column, dual thermal conductivity detector (TCD) chromatograph. One detector used a helium carrier with a Porapak Q column, and the other used an argon carrier with a molecular sieve column. Data reduction was aided by an Auto Lab System IV digital integrator equipped with a calculation module.

Analytical Analyses. The potassium remaining in the coal ash was determined with a Perkin-Elmer model 303 atomic absorption spectrophotometer after performing a J. Lawrence Smith ignition on the sample. To obtain a total potassium balance it was necessary to recover the potassium that adhered to the nickel catalyst by digesting the catalyst with acid and determining the potassium by atomic absorption. The amount of carbonate in the ash was determined by treating the ash with 1:1 HCl solution. The evolved gases were scrubbed, and the CO_2 was absorbed in Ascarite.

To determine the amount of potassium that could be readily extracted, ambient temperature and warm water washes were used for a designated time. The amount of potassium in the filtrate was determined by atomic absorption. Filtrate-evaporation was carried out, and the residue was analyzed by x-ray diffraction to determine the predominant form of the potassium compound.

X-Ray Diffraction. Ash and potassium carbonate pulverized to −325 mesh were examined with a General Electric XRD-5 diffractometer. Copper radiation at 35 kvp and 16 ma was used for the analyses. Each scan was started at an angle 2θ of 4° and continued through 70°. The data were recorded on a strip chart. The interplanar d spacings in angstroms for the recorded x-ray peaks were determined from a copper $K\alpha (\lambda = 1.5418 \text{ A})$ table, and compounds were identified from the ASTM x-ray powder diffraction file.

Methodology. All experiments were carried out by charging the reactor with the coal, potassium carbonate, and nickel catalyst mixture in a glove box under a nitrogen atmosphere. If nickel catalyst was not involved, the reactor was not charged in the glove box. In all other runs 100 grams of coal and about 110 grams of nickel catalyst with varying amounts of potassium carbonate were charged. The temperatures on the reactor and super heater were then brought to operating temperatures of approximately 650°C in less than 2 hours and were maintained at this value for the duration of the run. In all cases the run was 7 1/2 hours, and approximately 28 ml of water were added. The product gas was monitored for composition every half hour. The reactor pressure was *ca.* 30 psia, which is the pressure required to provide an adequate sample to the chromatograph. Runs were repeated until near duplication of two were obtained. The criteria for accepting the duplicate runs were based on the total mass balance and the quality of the gas produced. A total mass balance of ±4% was sufficient to ensure that there were no major leaks to prejudice the results. To ensure that the nickel catalyst was properly reduced and had not oxidized during loading in the mixed catalyst runs, only the runs in which the average gas had a heating value of over 800 Btu/scf (CO_2-free) were used. There were only two cases in a total of 12 separate runs with the mixed catalyst where the quality of the gas produced was below 800 Btu/scf. In both cases the runs were repeated a third time, and the criteria for acceptance was met. The amounts of methane and total gas produced per weight of coal are reported as volume rated average compositions.

Reactor Design. Coal gasification was carried out in a 1-inch od semicontinuous flow reactor described in earlier papers (8, 9). A flow diagram of this unit is shown in Figure 1. Because of the extremely active nature of the re-reduced nickel catalyst toward oxygen, it was stored under a nitrogen atmosphere. In addition, the reactor was charged under a nitrogen atmosphere in a glove box.

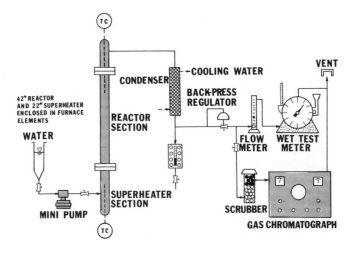

Figure 1. Schematic of 1-inch reactor

Results and Discussion

The minimization of the energy required to produce synthetic natural gas (SNG) from coal is of extreme interest in current process development. Since methane formation *via* reactions between hydrogen, coal, and carbon oxides is exothermic, the energy requirement for the endothermic coal–steam reaction will be decreased in proportion to the methane concurrently formed. Hence, the basic approach to this problem has been and continues to be maximizing the methane production in the first stage of the multiple step process by optimizing extensive and intensive conditions. Additional processing required for an acceptable SNG include adjustment of the H_2-to-CO ratio, scrubbing sulfur gases and carbon dioxide, and catalytic methanation. In contrast to this approach the NRRI coal conversion process uses catalysts to produce high methane content gas directly from coal–steam reactions in a single-stage reactor. Through such a conversion the heat requirement would be decreased and additional processing would be eliminated.

The principal chemical reactions intimately associated with the production of SNG from coal are Reactions 1–4. The goal of the NRRI coal conversion process is to carry out these reactions simultaneously in a single-stage reactor with suitable catalysts. Thus Reaction 5, which can be reached through a proper combination of Reactions 1–4, is indicative of the overall conversion.

$$C + H_2O = CO + H_2 \tag{1}$$

$$CO + H_2O = H_2 + CO_2 \tag{2}$$

$$CO + 3H_2 = CH_4 + H_2O \tag{3}$$

$$C + 2H_2 = CH_4 \tag{4}$$

$$C + H_2O = \tfrac{1}{2}CH_4 + \tfrac{1}{2}CO_2 \tag{5}$$

Table II compares non-, single-, and multiple-catalyzed coal–steam conversions in the previously described single-stage reactor. In spite of the difficulty in obtaining reproducible results and satisfactory mass balances on this small scale, the results are believed to represent conversions under the indicated conditions. Thus, there are several noticeable differences between the non-, single-, and multiple-catalyzed runs. They include coal conversion, product composition, total gas produced, product heating value, and gasification rates.

The per cent coal conversion increases when K_2CO_3 is used (*cf.* run 628 with 647 and 602). The addition of a nickel catalyst appears to have

Table II. Catalyzed and Non-Catalyzed Results

Run	628	647	614	602
Coal, grams	100	100	100	100
Catalysts, grams				
K_2CO_3	0	20	0	20
nickel	0	0	115	111
Avg. temp., °C	635	621	621	613
Pressure, psia	31	31	32	32
Time, hours	7.5	7.5	7.5	7.5
Coal conversion, %	50	60	51	56
Total gas, scf/ton	12270	17320	14470	19570
Gas composition, mole %				
H_2	40.2	32.1	17.6	13.0
CO	13.4	31.6	1.7	1.5
CO_2	32.0	22.2	39.9	39.4
CH_4	13.2	12.9	40.8	46.0
C_2H_6	0.7	0.3	0.0	0.0
unsats.	0.5	0.8	0.0	0.0
Bbl liq./ton	0.3	0.2	0.0	0.0
Scf H_2/ton	4930	5560	2550	2540
Scf CH_4/ton	1700	2773	5900	9000
Btu/scf (CO_2-free)	474	454	792	848

no effect on the conversion (*cf.* run 628 with 614). These results are consistent with the catalysis of Reaction 1 by the potassium carbonate.

The independent and combined influence of potassium carbonate and nickel on the quantity and distribution of products are also apparent from the data in Table II. The addition of K_2CO_3 decreases the quantity of liquid hydrocarbons produced over the non-catalyzed run. In the presence of the nickel catalyst or nickel and alkali catalysts (multiple catalyst) no liquids are observed. These observations indicate that the nickel catalyst gasifies and hydrocracks the liquid products that are produced in its absence. This accounts for the greater quantities of gas produced with the multiple catalyst than with the K_2CO_3 despite the volume contraction associated with methanation.

When K_2CO_3 is added, not only does the quantity of gas produced increase markedly but also does its composition. A significant increase in the per cent CO accompanied by a decrease in H_2 and CO_2 is observed. However, there is essentially no change in the CO_2-free heating value of the gas product. On the other hand, the run with the nickel catalyst shows a large increase in the product heating value, and this is reflected in the gas composition. The hydrogen and carbon monoxide decrease significantly, and the methane and carbon dioxide content increase. This is consistent with the methanation of the carbon monoxide by the nickel catalyst according to Reaction 3. The nickel catalyst also eliminates the C_2H_6 and unsaturates that are observed in its absence. The combined

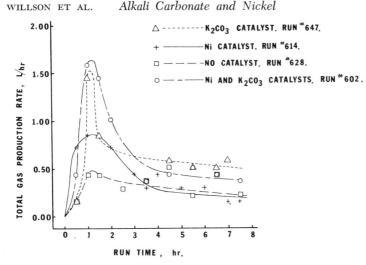

Figure 2. Comparison of gasification rates for non-, single-, and multiple-catalyst systems

effect of the nickel and alkali on the heating value and product gas composition is shown by run 602. Here the catalytic effect of the K_2CO_3 in the coal–steam reaction is combined with the hydrocracking and methanating function of the nickel catalyst to give a high Btu product gas and significant coal conversions.

The catalysts also influence the gasification rate. Figure 2 shows the rate of gas production which is time dependent. The rate reaches a maximum early in the run and then decreases to a lower, more constant rate. This early, rapid rate is attributed largely to the evolution of volatiles from the coal. This type of decrease in gasification rate with con-

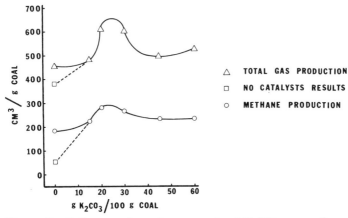

Figure 3. Influence of varying amounts of K_2CO_3 on methane and total gas production

Table III. Influence of K_2CO_3

Run	628[b]	614	607
Catalysts			
K_2CO_3, grams	0	0	15
Ni, grams	0	115	111
Coal conversion, %	50	51	55
Material balances, %			
total	103	98	99
potassium	—	—	94
carbonate	—	—	98
cm$_3$ CH$_4$/gram coal	51	185	225
cm$_3$ gas/gram coal	382	453	481
Gas composition, mole %			
H$_2$	40.2	17.6	12.5
CO	13.4	1.7	0.7
CO$_2$	32.0	39.9	40.1
CH$_4$	13.2	40.8	46.7
C$_2$H$_6$	0.7	0.0	0.0
unsats.	0.5	0.0	0.0
Btu/scf (CO$_2$-free)	474	792	861

[a] All runs were made with 100 grams of coal over 7½ hours at about ½ atm and 650°C.

version (attributed to a decrease in the reactivity of coal with increased carbon burn off) is typical for coal. The runs with K_2CO_3 have the greatest gasification rates. Although the run using only nickel has a lower rate than either of the runs using K_2CO_3, it is still significantly greater than the run with no catalyst. This is largely the result of the gasification and hydrocracking of the liquids that are normally present in the absence of the nickel catalyst.

The results of the investigation on optimizing the quantity of potassium carbonate in the multiple catalyst system are in Table III. The methodology used is that presented in the Experimental section. Furthermore, the results previously discussed are also applicable to this data. Hence, the catalytic effect on the product gas composition as reflected in its heating value is observed. Also the maximum methane and total gas yielded were produced with 20 grams of K_2CO_3 in the system. Significant increases or decreases of the K_2CO_3 content from this amount lowers the amount of methane and gas produced. These results are presented in Figure 3. The dashed lines indicates the results where neither alkali nor nickel catalyst was present. The optimum methane and total gas coincide with 20–25 grams of K_2CO_3 per 100 grams of coal.

The relatively constant product heating value, in spite of the changes in the potassium content, is indicative of the nickel catalyst's methanating function. This agrees with previously presented results. On the other on Coal–Steam Gasification[a]

602	631	608	611
20	30	45	60
111	111	113	114
56	56	60	60
103	99	96	98
106	106	95	99
101	104	102	105
281	266	232	234
612	603	493	524
13.0	12.6	12.1	13.5
1.5	2.6	1.1	1.2
39.4	40.7	40.6	40.6
46.0	44.1	46.2	44.7
0.0	0.0	0.0	0.0
0.0	0.0	0.0	0.0
848	837	860	843

[b] This run also produced about 4 ml of liquid hydrocarbons.

hand, the gas production appears to depend heavily on both catalysts. As previously mentioned, this is associated with the nickel catalyst's ability to gasify the liquid hydrocarbons produced and the catalytic influence of K_2CO_3 on the coal–steam reaction.

The potassium carbonate reagent and ash samples were examined by x-ray diffraction to identify the forms of potassium carbonate and other potassium compounds present. Only $K_2CO_3 \cdot 1\frac{1}{2}H_2O$ and K_2CO_3 were detected in both materials. The presence of the same potassium compounds in the ashes that were in the original reactor charges would indicate that the forms of potassium do not seem to have been changed by the coal conversion reactions. However, some of the potassium carbonate could have been converted to amorphous potassium compounds or compounds of insufficient quantities to be detected by x-ray diffraction. Amorphous compounds usually are not identifiable by x-ray diffraction.

The relative amounts of potassium carbonate in the ashes were determined by comparing the intensities of the x-ray diffraction peaks. Intensities are proportional to the amount of a given compound. Ash samples containing 0, 15, 20, 45, and 60 grams of potassium carbonate in the original reactor charges were examined. The sample with no potassium carbonate showed SiO_2 and one or more unidentifiable ash constituents. This observation is consistent with the chemical composition of the coal. The 15-gram sample did not show potassium carbonate. This is perhaps an indication that some of the originally charged potassium carbonate

Table IV. Sulfur

Run	Sulfur In, grams		
	Catalyst	Coal	Total
651	.078	.698	.776
652	.082	.698	.780
658	.092	.698	.790
671	.078	.680	.758
675	.078	.680	.758
680	.073	.680	.753
686	.063	.680	.742
687	.065	.667	.733
688	.068	.667	.735
689	.066	.567	.633

may have been converted to an amorphous potassium compound. On the other hand, the 20-, 45-, and 60-gram samples show $K_2CO_3 \cdot 1\frac{1}{2}H_2O$ and K_2CO_3 in amounts proportional to those in the original reactor charges. If significant quantities of additional potassium compounds were formed, they should be apparent in the diffraction pattern of the samples using the larger amounts of potassium.

Wet chemical analyses of the potassium carbonate were done to support the x-ray diffraction investigation. These analyses also permitted material balances to be carried out on this compound as demonstrated by the material balance data in Table III. The analytical data support the x-ray diffraction data in that the alkali carbonate is the predominant form of the potassium compound in the coal ash. Furthermore, for runs using 15–60 grams K_2CO_3, between 80–100% of the potassium was retained in the ash. The average potassium retained in the ash for 10 samples was 88 wt %. The remaining potassium either involved a loss to the experimental system or was unaccounted for because of sampling difficulties.

To address the recoverability of the alkali carbonate catalyst from the ash, both ambient and warm water washes were used. In the ambient temperature wash, 1 gram of ash was stirred 1 hr in 100 ml H_2O. Between 80–100% (87% average for 24 samples) of the initial potassium used in the gasification run could be dissolved in the aqueous solution. The range of potassium solubility reflects some of the difficulties previously referred to. No attempt was made to recrystallize potassium compounds from the ambient temperature washes.

In the warm water washes (65°–75°C) for potassium compound solubility 50 grams of ash were stirred for ½ hour in 80 ml H_2O. Only about 75% of the potassium was dissolved; the remaining 25% remained

Mass Balances

	Sulfur Out, grams		Original Coal Sulfur In, %			
Catalyst	Ash	Total	Catalyst	Ash	System Loss	% Conv.
.363	.355	.718	40.7	50.8	8.5	62
.282	.452	.734	28.7	64.8	6.5	61
.250	.515	.765	22.8	73.8	3.4	62
.277	.442	.720	29.2	65.0	5.8	55
.397	.232	.628	46.9	34.1	19.0	72
.304	.320	.624	33.9	47.1	19.0	64
.335	.226	.561	40.0	33.2	26.8	50
.360	.322	.682	44.2	48.3	7.5	59
.338	.312	.650	40.5	46.8	12.7	76
.308	.201	.509	42.6	35.4	22.0	60

in the ash. Evaporation of the filtrate to dryness produced a crystalline compound. X-ray diffraction revealed K_2CO_3 and $K_2CO_3 \cdot 1\frac{1}{2}H_2O$. Further analysis revealed 50 wt % potassium and 25 wt % CO_3 with the remaining 20% being attributed largely to hydrated water.

The ash residue from the warm water extraction was examined by x-ray diffraction. Although there was no K_2CO_3 or $K_2CO_3 \cdot 1\frac{1}{2}H_2O$ detected in this ash, $CaCO_3$ was evident. The retention of the calcium carbonate and dissolution of the potassium carbonates with water are consistent with the solubility of these compounds in aqueous solutions.

Since sulfur gases poison nickel methanation catalysts, sulfur material balances were done to determine the extent of sulfur buildup on the catalyst. This was accomplished by determining the total sulfur in the methanation catalyst and coal charged to the reactor and again on the ash and catalyst removed from the reactor. Representative results for 10 independent runs under similar experimental conditions are in Table IV. These data demonstrate that some of the sulfur is lost to the system, some remains in the ash, and some reacts with the catalyst. Furthermore, the considerable variance of the data for the independent runs indicates that the fate of the sulfur is very sensitive to experimental conditions.

The per cent of the coal's original sulfur that has been deposited on the nickel methanation catalyst is shown in Table IV; it amounts to between 23 and 47%. Although x-ray diffraction failed to reveal the combined form of this sulfur, probably because of its small concentration, it is presumed to be nickel sulfide formed by Reaction 6:

$$2H_2S + 3Ni = Ni_3S_2 + 2H_2 \tag{6}$$

The presence of the sulfide in place of the sulfate is in agreement with the reducing atmosphere within the system and has been supported by qualitative analyses performed on the catalyst. With these small buildups in sulfur, no decrease in the catalyst activity was observed under experimental conditions. This is likely the result of the large amount of catalyst used. By using an average value of 37% of the coal's sulfur going to the catalyst from these runs one can calculate the amount of coal required completely to sulfide the nickel methanation catalyst which contains 35% nickel. This value is 42.4 grams of coal per gram of catalyst or 46.0 lbs of catalyst per ton of coal. In view of the price of the nickel methanation catalyst and provided such a data extrapolation is valid, such a high catalyst usage would be economically unfeasible without some means of catalyst regeneration.

Table IV also shows the per cent of the coal's sulfur that remains in the ash. Preliminary results have indicated that about 70% of the sulfur remaining in the ash under these conversion conditions is in the sulfate form. This represents a considerable increase in the sulfate form since the coal originally contained only about 0.05% sulfur in this form. Although the sulfate compounds in the ash have not been identified, the Glenrock coal is known to contain considerable amounts of Ca, Mg, Fe, and Al, all of which could form sulfates.

The per cent of the coal's sulfur that is lost to the system has been included in Table IV. This loss has been attributed to deposition on the reactor walls, dissolution in the liquid product, and evolution from the system as a gaseous compound. No attempt has been made to pursue these latter areas although the reactor scale contains sulfur as a sulfide.

Conclusions

A high heating value product gas ($\sim$850 Btu/scf, CO_2-free) can be produced directly from coal–steam reactions using a single-stage reactor in conjunction with a multiple catalyst. The conversion ($\sim$60%) is carried out at 2 atm and 650°C. The multiple catalyst consists of potassium carbonate and a nickel methanation catalyst. The influence of each catalyst on the coal–steam reactions is combined in the integrated system. Potassium carbonate increases the total gas production and rate while the nickel catalyst hydrocracks the evolved liquids and methanates the carbon oxides.

In the presence of the methanation catalyst the K_2CO_3 : coal ratio was optimized at about 20–25 wt %. This amount of the alkali carbonate has produced the maximum total gas and methane production under experimental conditions. A deviation from this quantity of potassium carbonate was reflected in a decrease in both the total gas and methane production, particularly at lower concentrations.

X-ray diffraction and wet chemical analysis of the ash from the multiple catalyst runs revealed the presence of potassium carbonate and its 1½ hydrate. This was the initial form of the potassium charged to the system. Material balances on the potassium carbonate showed that although the majority of this compound ends up in the ash, some of it is lost to the system.

The extraction of the ash with single water washes revealed that significant quantities of the potassium could be easily dissolved (generally in excess of 80%). This agrees with the identification of K_2CO_3 and $K_2CO_3 \cdot 1½H_2O$ as the predominant potassium compounds that are extremely soluble in water. By evaporating the water wash to dryness, crystalline hydrated and anhydrous potassium carbonate were obtained. This recovered product contained about 75% of the potassium initially in the ash.

Sulfur material balances on the system showed that the sulfur contained in the coal charge became primarily distributed between the ash and methanation catalyst while some was lost to the system. The buildup of sulfide sulfur on the nickel catalyst amounted to 25–45% of that initially contained in the coal. This acquisition of sulfur by the methanation catalyst indicates that periodic catalyst regeneration would probably be necessary to maintain it in a sufficiently active form. In spite of the coal's predominant form of sulfur being organic, the sulfur remaining in the ash was predominantly in the sulfate form.

Acknowledgments

This work was sponsored by the U. S. Department of Interior, Office of Coal Research and is presented with their permission. The technical assistance and unwavering enthusiasm provided by E. J. Hoffman, W. E. Duncan, and J. A. Roberts are greatly appreciated.

Literature Cited

1. Fox, D. A., White, A. H., *Ind. Eng. Chem.* (1931) **23**, 259–266.
2. Taylor, H. S., Neville, H. A., *J. Amer. Chem. Soc.* (1921) **43**, 2055–2071.
3. Kröger, C., *Angew. Chem.* (1939) **52**, 129–139.
4. Lewis, W. K., Gilliland, E. R., Hipkin, H., *Ind. Eng. Chem.* (1953) **45**, 1697–1703.
5. Lefrancois, P. A., Barclay, K. M., Skaperdas, G. T., ADVAN. CHEM. SER. (1967) **69**, 64–80.
6. Storch, H. H., "Chemistry of Coal Utilization; Synthesis of Hydrocarbons from Water Gas," Vol. 2, Chap. 39, pp. 1797–1800, Wiley, New York, 1945.

7. *Ibid.*, pp. 1818–1824.
8. Hoffman, E. J., *Amer. Chem. Soc., Div. Petrol. Chem., Prepr.* (1971) **16** (2), C 20.
9. Hoffman, E. J., Cox, J. L., Hoffman, R. W., Roberts, J. A., Wilson, W. G., *Amer. Chem. Soc., Div. Fuel Chem., Prepr.* (1972) **16** (2), 64–67.

RECEIVED May 25, 1973.

13

Catalyzed Hydrogasification of Coal Chars

N. GARDNER, E. SAMUELS, and K. WILKS

Case Western Reserve University, Cleveland, Ohio 44106
Cleveland, Ohio 44106

A kinetic study of catalyzed and non-catalyzed coal char hydrogasification was accomplished using a thermobalance (0–1000°C, 0–1000 psi). Rate data were correlated with a kinetic model in which the activation enthalpy was presumed to be a linear function of extent of reaction. Deposition of catalysts ($KHCO_3$, K_2CO_3, and $ZnCl_2$) on the char and subsequent gasification resulted in substantially increased reaction rates. The effectiveness of the catalysts were in the order $KHCO_3 \approx K_2CO_3 > ZnCl_2$. Electron microprobe and scanning electron microscopy of chars revealed good catalyst distribution throughout the char particles.

The reaction of hydrogen with coal and coal chars to produce gaseous hydrocarbons (hydrogasification) has received considerable attention for at least 35 years since Dent et al. in 1937 first reported on the hydrogasification synthesis (1). The reaction proceeds in two steps. In the initial stage, reaction rates are extremely rapid as the volatile matter and more reactive components of the coal are gasified. Subsequent rapid hydrogenolysis of the higher homologs formed yields methane. In the second stage of the reaction the structure of the remaining carbon char is more graphitic in character, resulting in a much slower hydrogasification rate. Here we report on the catalysis of the slow, second stage of the hydrogasification reaction.

There have been numerous reports and patents on the catalysis of a similar reaction—the liquid-phase hydrogenation of coal to liquid and gaseous products. Hydrogenation reactions are generally performed at several hundred atmospheres and at 400°–500°C where the hydrocarbon products formed are substantially liquid. The ability of tin-halogen compounds, ammonium molybdate, and many other materials to catalyze the coal hydrogenation reactions is well known (2). Although the reaction is carried out under conditions where coal has undergone agglomeration and liquefaction, the method of contacting catalyst and coal particles has a strong influence on reaction rate. For example, the addition of powdered ferrous sulfate to coal particles has almost no effect on the hydrogenation rate (3). Impregnation of the coal by immersing it in aqueous solutions of ferrous sulfate and following by oven drying resulted in a sharp increase in hydrogenation rate with high productions of asphalt and oil. Impregnated nickelous chloride, stannous chloride, and ammonium molybdate show similar increases in catalytic activity compared with powders of the same materials (4).

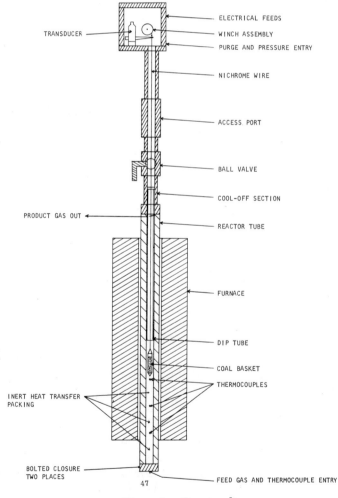

Figure 1. Reactor diagram

There have been extensive studies on the ability of particulate metals and metal salts to catalyze the reactions of graphitic carbon with oxygen and carbon dioxide (see Ref. 5 for an excellent review). For example, colloidal iron on Ticonderoga graphite reduces the activation energy for the carbon–oxygen reaction from 46 to 10 kcal/mole. A 7% iron deposit impregnated from solution on sugar char reduced the activation energy from 61.2 to 22.8 kcal/mole for the carbon–carbon dioxide reaction. In addition, dispersions of metals in carbon have been prepared by carbonization of polymers containing metal salts. The dispersions are catalytically active in the gasification reactions with carbon dioxide and oxygen. The mechanism of the substantial reduction in activation energy is not clear although much quantitative information has been obtained. Two types of mechanisms have been proposed, oxygen transfer and electron transfer. In the oxygen-transfer mechanism the catalyst is presumed to assist the dissociation of molecular oxygen to chemisorbed atomic oxygen which then reacts with the carbon surface. Electron-transfer mechanisms involve the pi electrons of graphitic carbon and the vacant orbitals of the metal catalysts. The catalytic effect presumably results from the altered electronic structure of the surface carbon atoms.

In contrast to hydrogenation and oxidation reactions, much less is known about the ability of materials to effect the catalysis of hydrogasification reactions. Alkali carbonates, 1-10 wt % catalyze the hydrogasification of coals and cokes at 800°–900°C (6). The suggested mechanism is that adsorption of the alkalies by carbon prevents graphitization of the surface. Zinc and tin halides are effective hydrogasification catalysts. There is, however, little kinetic information on any of the catalyzed hydrogasification reactions.

This kinetic study of catalyzed hydrogasification reactions utilizes a high temperature, high pressure recording balance. A thermobalance is particularly useful in gas–solid reactions because the weight of small solid samples can be measured continuously. Direct kinetic analysis of the weight loss curves are straightforward.

Equipment and Procedures

The high pressure thermobalance is very simliar to the balance described by Feldkirchner and Johnson (7). The thermobalance is designed to operate isothermally up to 1000°C and 2000 psi hydrogen. Details of the balance are shown in Figure 1, and a schematic of the system is shown in Figure 2. The reactor tube is constructed of Haynes 25 superalloy. The mass transducer is a Statham model UC 3 attached to a balance arm (Micro-scale accessory UL 5) and has a full-scale range of 6 grams.

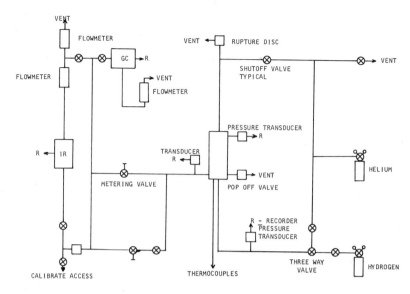

Figure 2. Schematic of coal gasification system

The sample is lowered into the reaction zone by an electric motor-driven windlass at about 1/2 inch/sec. The position of the sample in the reactor is obtained by monitoring the output of a small 10-turn potentiometer which is coupled directly to the windlass. Temperatures in the reactor are measured by stainless steel-encased chromel–alumel thermocouples, the closest one to the sample being located 1/4 inch below the sample. Hydrogen flow rates are controlled ±5% over the range 10–40 scf/hr. Gas analysis is obtained by splitting a portion of the gas product stream to an infrared detector where methane content is continuously measured and a portion to a gas chromatograph where total gas composition is determined.

Table I. Chemical Analysis

Component	Char A (Hydrogen Pretreated)	Char B (Oxygen Pretreated)
	Weight Per Cent	
Carbon	81.45	69.69
Hydrogen	1.46	4.39
Oxygen	3.76	12.47
Ash	13.94	10.31
Total	100.61	96.86

All experiments used chars supplied by the Institute of Gas Technology. Char A was hydrogen-pretreated Pittsburgh Seam, Ireland Mine bituminous coal. Char B was also prepared from a Pittsburgh Seam, Ireland Mine coal pretreated (about 1 ft^3 of oxygen/lb of fresh coal at

400°C) in an air-fluidized bed. An analysis of the two chars is shown in Table I. The char was sized 18 × 35-mesh sieve fraction. The sample weight in any given run was 1.5-2.5 grams. The sample bucket was constructed of 100-mesh stainless steel screen.

Catalysts were deposited on the char particles by evaporation from solution. Catalyst concentrations were 5 wt % metal. Catalyst distribution on the char was examined by electron microprobe and scanning electron microscopy.

Results and Discussion

Reaction of Non-Catalyzed Chars. Initial runs were performed on both chars to determine non-catalyzed reaction rates. The fractional conversion of the char, defined as

$$X = 1 - \frac{\text{wt of char at time } t - \text{wt ash}}{\text{wt of char initially} - \text{wt ash}}$$

is shown as a function of time for chars A and B, respectively, in Figures 3 and 4.

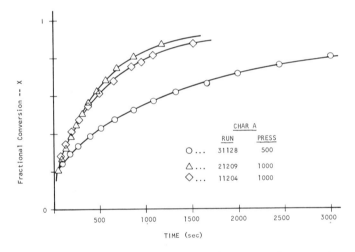

Figure 3. Non-catalyzed hydrogasification of Char A at 500 and 1000 psi, 950°C

Characteristically, the fractional conversion curves show high initial reaction rates as the more volatile matter in the char is gasified followed by a much slower reaction regime where the rate slowly diminishes as the char is consumed. Such phenomena have been described by several investigators (8–12).

For kinetic analysis of the weight-loss data we propose a model different from those previously discussed. We assume that the reaction rate is given by the following kinetic expression:

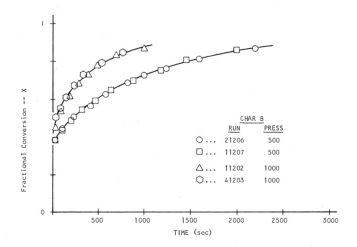

Figure 4. Non-catalyzed hydrogasification of Char B at 500 and 1000 psi, 950°C

$$\frac{dX}{dt} = k\, P^n_{H_2} (1-X) \exp[-\Delta H^{\neq} RT] \qquad (1)$$

where X = fractional conversion of char
k = frequency factor
n = order of reaction
$\Delta H^{\neq}$ = activation enthalpy for gasification in kcal/mole
P_{H_2} = hydrogen pressure in atm

In contrast to homogeneous reactions, where activation enthalpies are independent of the extent of reaction, hydrogasification activation enthalpies are clearly a function of the extent of reaction. One mechanism postulated by a number of investigators is based on the carbon structure's becoming more graphitic with increasing reaction. In the absence of any other information, the simplest function for $\Delta H^{\neq}(X)$ is a linear form

$$\Delta H^{\neq}(X) = \Delta H^0 + \alpha X \qquad (2)$$

where ΔH^0 = initial activation enthalpy
α = factor that determines sensitivity of $\Delta H^{\neq}$ to X

Substituting this expression into Equation 1 yields

$$\frac{dX}{dt} = kP^n_{H_2}(1-X)\exp(-\Delta H^0/RT)\exp(-\alpha X/RT) \qquad (3)$$

By combining the pressure and temperature terms into two constants, Equation 3 can be simplified to test its applicability as a rate expression. Thus,

$$\frac{dX}{dt} = kP^n{}_{H_2} \exp(-\Delta H^\circ/RT) \cdot \exp\left(\frac{-\alpha}{RT} \cdot X\right) \cdot (1 - X)$$

and then

$$\frac{dX}{dt} = K \exp(-bX) \cdot (1 - X) \tag{4}$$

where K and b are both constants and equal to

$$K = kP^n{}_{H_2} \exp(-\Delta H^\circ/RT)$$
$$b = \alpha/RT$$

Rearrangement and integration of Equation 4 gives the final form of the rate expression used to test the kinetic data.

$$\int_0^x \frac{\exp(bX)}{(1-X)} dX = \int_0^t K\,dt = Kt \tag{5}$$

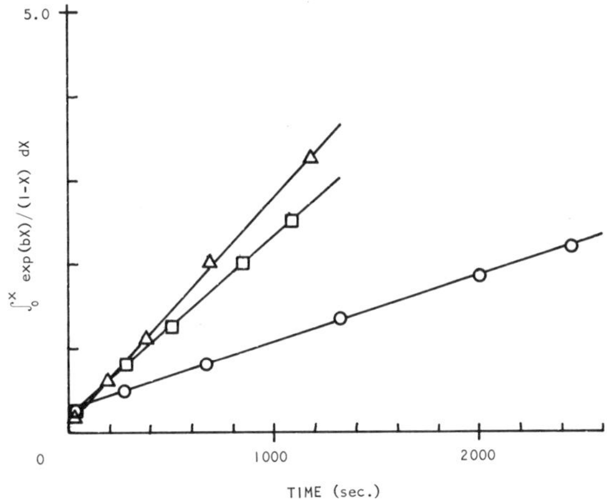

RUN NO.	PRESSURE (psia)	b	K (1/sec.)
○...31128	500	.9	.00081
□...11204	1000	.8	.00212
△...21209	1000	.8	.00274

Figure 5. Kinetic test for non-catalyzed hydrogasification of Char A at 500 and 1000 psia, 950°C

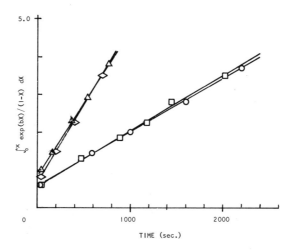

RUN NO.	PRESSURE (psia)	b	K (1/sec.)
O...21206	500	1	.00143
□...11207	500	1	.00147
◇...11202	1000	1.2	.00406
△...41203	1000	1.2	.00386

Figure 6. Kinetic test for non-catalyzed hydrogasification of Char B at 500 and 1000 psia, 950°C

Table II. Tabulation of b and K Values for Non-Catalyzed Char A and B

Char A (Hydrogen Pretreated)

500 psi H_2			1000 psi H_2		
Run	b	K	Run	b	K
11122	0.25	0.000974	41121	0.8	0.00386
31128	0.9	0.000810	21209	0.8	0.00274
Average	0.57	0.000892	11204	0.8	0.00212
			Average	0.8	0.00291

Char B (oxygen Pretreated)

500 psi H_2			1000 psi H_2		
Run	b	K	Run	b	K
11124	1	0.000818	11127	0.9	0.00399
21206	1	0.00143	31209	0.8	0.00290
11207	1	0.00147	41203	1.2	0.00386
31207	0.5	0.00125	11202	1.2	0.00406
Average	0.88	0.00124	Average	1.025	0.00372

Figures 5 and 6 show plots of the integral on the left side of Equation 5 vs. time for chars A and B, respectively. The parameter b is chosen to minimize the sum of the squares of the errors of a last-square fit to a straight line through the data points. The value of K is then evaluated from the slope of the straight line.

Table II shows the values of b and K determined from Figures 5 and 6. The values of b are independent of hydrogen pressure within experimental error. The ratio of the K values at 1000 and 500 psi, respectively,

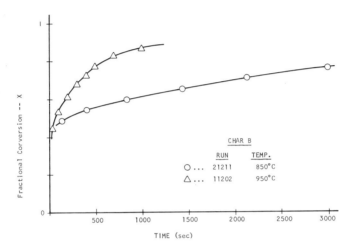

Figure 7. Effect of temperature on the non-catalyzed hydrogasification of Char B at 1000 psi

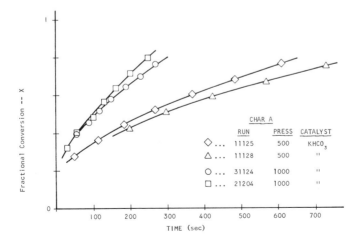

Figure 8. Catalyzed hydrogasification of Char A at 500 and 1000 psi, 950°C

226   COAL GASIFICATION

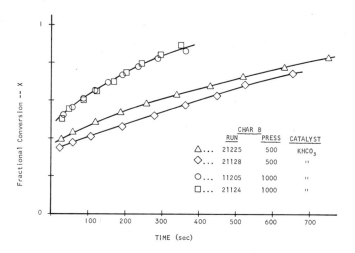

Figure 9. Catalyzed hydrogasification of Char B at 500 and 1000 psi, 950°C

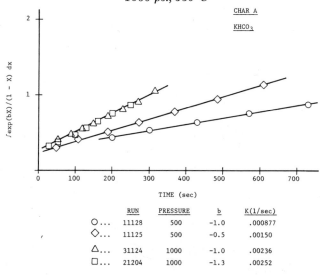

Figure 10. Kinetic test for $KHCO_3$-catalyzed hydrogasification of Char A

are for char A, $K_{1000}/K_{500} = 3.27$ and for char B, $K_{1000}/K_{500} = 2.89$. The hydrogen pressure appears in the K term raised to the power n, where n is the order of the reaction. Thus, the hydrogasification reaction order is approximately 1.6 for char A and 1.5 for char B. Notwithstanding the scatter in the data, it appears that the hydrogen order is $n = 3/2$.

The y-axis intercept in Figures 5 and 6 should have been zero. The positive non-zero intercept results from the very rapid first stage of the reaction. For the oxygen-pretreated char B, the amount of carbon gasified in the rapid first stage of the reaction is greater than that for the hydrogen-pretreated char. This effect is caused by the pretreatment of the char which results in a high per cent of volatile matter. In the second, slower part of the reaction, the chars behaved almost identically as indicated by the similar b and K values.

Figure 7 shows the fractional conversion as a function of time for char B at 850° and 950°C, 1000 psi. The values of b and K at 850°C were 2.3 and 0.00128 sec^{-1}, respectively. From the variation of K with temperature, ΔH^0 can be estimated to be 29.3 kcal/mole. The activation enthalpy is then given by:

$$\Delta H^{\neq} \text{ (kcal/mole)} = 29.3 + 2.43X \qquad (6)$$

for the uncatalyzed char B system.

Reaction of Catalyzed Chars. In this preliminary study we have concentrated on catalysts that are known to accelerate the hydrogasification reaction—the alkali metals and zinc salts (6). Figures 8 and 9 show the fractional conversion vs. time data for a $KHCO_3$ catalyst deposited on chars A and B; a substantial catalytic effect is found. The time to achieve a gasification fraction X is roughly halved by the $KHCO_3$ catalyst. The kinetic analysis for the parameters b and K is shown in Figures 10 and 11, where the kinetic equation fits the data to high conversions.

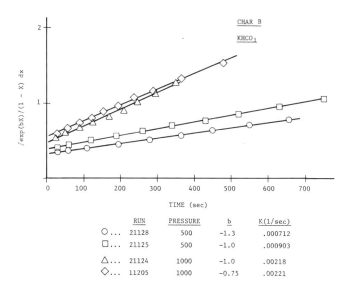

Figure 11. Kinetic test for $KHCO_3$-catalyzed hydrogasification of Char B

Table III. Tabulation of b and K Values for Char A and B Catalyzed with $KHCO_3$

Char A (Hydrogen Pretreated)

500 psi H_2			1000 psi H_2		
Run	b	K	Run	b	K
11128	−1.0	0.000877	11208	−0.3	0.00224
11125	−0.5	0.0015	21204	−1.3	0.00252
Average	−0.75	0.00119	31124	−1.0	0.00236
			Average	−0.867	0.00237

Char B (Oxygen Pretreated)

500 psi H_2			1000 psi H_2		
Run	b	K	Run	b	K
21128	−1.3	0.000712	21124	−1.0	0.00218
21125	−1.0	0.000903	11205	−0.75	0.00221
Average	−1.15	0.000807	Average	−0.88	0.00219

Evaluation of the b and K parameters for $KHCO_3$-catalyzed reactions is shown in Table III.

The rate enhancement by the catalysts is shown by the evaluation of α from the parameter b. Taking −1 to be the average value of b for the char B system, the corresponding value of α is −2.43. This term makes the linear expression for the activation enthalpy decrease with increasing carbon gasification.

$$\Delta H^{\neq} \text{ (kcal/mole)} = 29.3 - 2.43X$$

Thus, the net effect of the catalyst is that the activation enthalpy decreases (*e.g.*, the reaction becomes easier) with increasing extent of reaction.

Potassium carbonate (K_2CO_3) and zinc chloride ($ZnCl_2$) were also studied. These catalysts were deposited by impregnation at a 5 wt % metal concentration. The K_2CO_3 catalyst behaved similarly to the $KHCO_3$. The weight-loss curves for this catalyst at 500 and 1000 psi, 950°C are shown in Figure 12. Kinetic analysis of this catalyst produced values of b and K similar to those for $KHCO_3$ (*see* Table IV).

Zinc chloride also showed a catalytic effect (Figure 13) but was not as effective as the potassium catalyst. Kinetic analysis of the rate data gave average values for b of 0.37 and 0.35 at 500 and 1000 psi, respectively, and average values for K of 0.0015 and 0.0039 (Table IV). Figure 13 also shows a direct comparison of relative effectiveness of the zinc and potassium salt catalysts.

Figure 14 shows a representative composite plot of the fractional conversion, the methane rate of production, and the temperature *vs.*

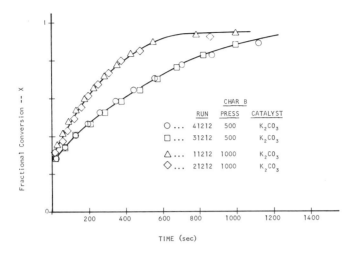

Figure 12. Catalyzed hydrogasification of Char B at 500 and 1000 psi, 950°C

Table IV. Tabulation of b and K Values for Char A and B Catalyzed with other Catalysts

Char A (Hydrogen Pretreated − $ZnCl_2$)

	500 psi H_2			1000 psi H_2	
Run	b	K	Run	b	K
21214	0.5	0.00160	21208	0.5	0.00492
31214	0.25	0.00140	52208	0.2	0.00301
Average	0.37	0.00150	Average	0.35	0.00397

Char B (Oxygen Pretreated − K_2CO_3)

	500 psi H_2			1000 psi H_2	
Run	b	K	Run	b	K
41212	−0.15	0.00155	21212	−0.5	0.00245
31212	−1.2	0.000867	11212	−0.5	0.00248
Average	−0.67	0.00118	Average	−0.5	0.00247

time for Char A with a $KHCO_3$ catalyst. The concentration of methane in the product stream is proportional to the rate of the hydrogasification reaction, as

$$r = \frac{dP_{CH_4}}{dt} = \frac{1}{12}\frac{1}{n_c}\left(\frac{dV}{dt}\right)\frac{P_{CH_4}}{RT} \quad (7)$$

where

r = rate (moles/min/initial grams of carbon)

n_c = initial moles carbon

P_{CH_4} = partial pressure CH_4 in atm

R = gas constant in liter atm/moles°K

T = temperature, °K

$\left(\dfrac{dV}{dt}\right)$ = product gas flow rate in liters/min

Equation 7 can be integrated numerically. The result is shown in Figure 15.

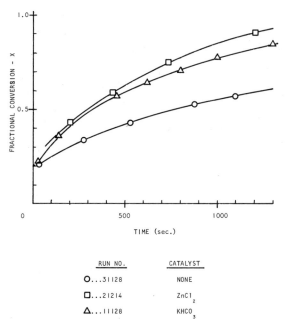

RUN NO.	CATALYST
○...31128	NONE
□...21214	$ZnCl_2$
△...11128	$KHCO_3$

Figure 13. Comparison of Char A-catalyst systems at 950°C and 500 psi

Although the infrared measurement of methane production leads to qualitative agreement with the direct mass determination, quantitative agreement is not good. This is most probably a result of axial dispersion in the gas product stream which results in a loss of kinetic information, difficulties in precisely regulating the product stream flow rate which would lead to cumulative errors, and the formation of small amounts of higher molecular weight hydrocarbons.

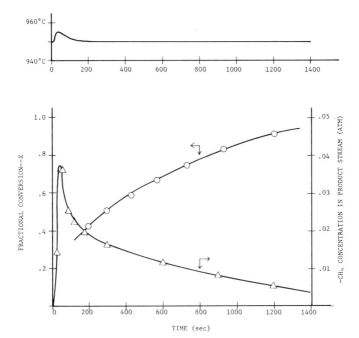

Figure 14. Composite plot of rate data for $KHCO_3$-catalyzed Char B, run 11128, 500 psi, and 950°C

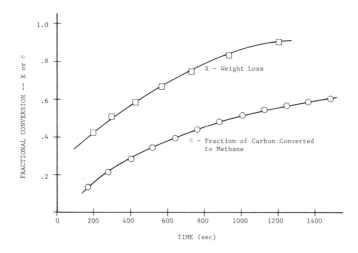

Figure 15. Fractional conversion of carbon to methane for $KHCO_3$-catalyzed char A at 500 psi, 950°C

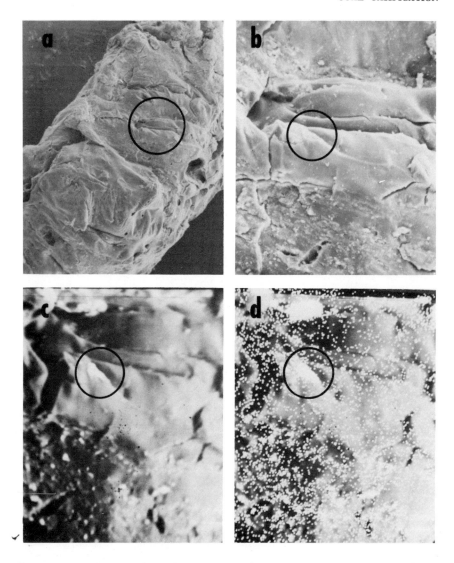

Figure 16. Micrographs of $KHCO_3$-catalyzed Char A. a, SEM—75×; b, SEM—300×; c, BSE—200×; d, iron x-ray on BSE—200×.

The temperature-*vs.*-time curve shown in Figure 14 indicates a small temperature increase resulting from the exothermic hydrogasification reaction. Unfortunately, the actual temperature of the char has not been measured.

Microscopy of the Chars. The chars were examined by scanning electron microscopy and an electron microprobe analyzer for particle structure, catalyst distribution, and structural changes at the catalyst sites as a result of gasification. The scanning electron microscope can take high magnification, high resolution pictures of the char particles while the microprobe analyzer can be calibrated to scan for any element on the char particle surface.

Presented here is a cross section of representative photographs for each char. Proper interpretation of these pictures is important. The

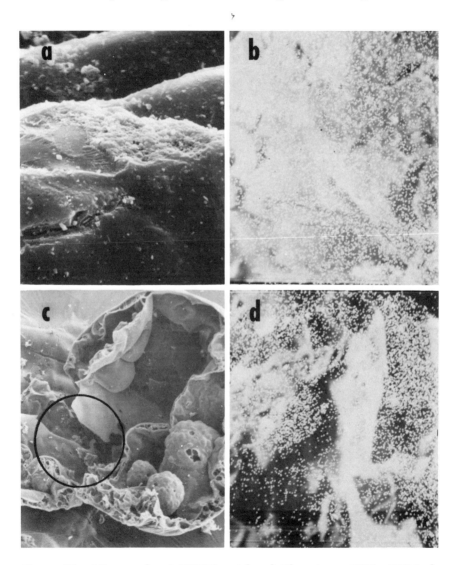

Figure 17. Micrographs of $KHCO_3$-catalyzed Char A. a, SEM—200×; b, potassium x-ray on BSE—200×; c, cut particle-SEM—100×; d, cut particle, potassium x-ray on BSE—200×.

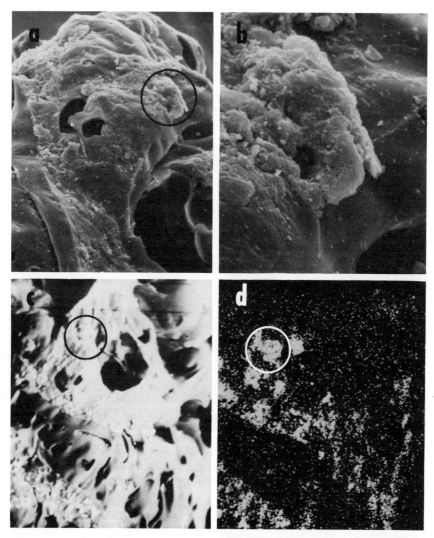

Figure 18. Micrographs of uncatalyzed Char B. a, SEM—500×; b, SEM—2000×; c, BSE—200×; d, iron x-ray—200×.

scanning electron micrographs (SEM) are high quality pictures but are often slightly distorted at low magnification (75–100×). The back-scattering electron micrographs (BSE) and x-ray micrographs are all taken on the electron microprobe analyzer. The BSE micrographs from this instrument are of poor clarity because the microprobe was designed for x-ray analysis. An x-ray micrograph is a scattering of white dots on the picture. If the element being analyzed is not present, there is an even but sparse distribution of white dots on the micrograph. This is an artifact of the analyzer. The locations where the element being analyzed is present are then represented by a concentration of white dots.

Figure 16a shows a low magnification scanning micrograph of a hydrogen-pretreated char A particle coated with 5% $KHCO_3$. Higher magnification micrographs of the center of this particle are shown in Figures 16b and 16c (see the area circled in Figure 16a). These are both of the same area on the particle. Figure 16b is a scanning micrograph while 16c is a BSE micrograph. Figure 16d is the iron x-ray superimposed on the BSE of Figure 16c. This shows areas of the particle surface where iron deposits or ash concentrations are located.

A close-up of the ash deposit circled in Figure 16b and 16c is shown in Figure 17a. Figure 17b shows a potassium x-ray superimposed on the BSE of Figure 16c. The heavy but even concentration of white dots indicates that the catalyst is present in a well distributed manner. The catalyst apparently exists on the particle surface in a finely divided state because there are no distinguishable clumps.

Micrographs of an uncatalyzed char B type particle that has undergone gasification for 8 min, losing about 50% of its carbon content, are shown in Figure 18. The area encircled in the scanning micrograph of Figure 18a is shown enlarged (2000×) in Figure 18b. This area has been identified as one particularly high in iron content by the electron microprobe analyzer and is probably part of a shell formed from the ash. Comparison of the back-scatter micrograph (Figure 18c) and the iron x-ray micrograph (Figure 18d) reveals that the porous, glassy-like portion of the micrograph is ash-free and predominantly carbon while the rough areas are predominantly ash.

None of the micrographs obtained yielded any information on structural changes at catalyst sites or near ash deposits. The principal information gleaned from the microscopy is that there is good catalyst distribution throughout the chars that have been pretreated in oxygen and hydrogen.

Acknowledgments

The thermobalance was made available by Consolidated Natural Gas Service Co., Inc. Special thanks go to Raymond R. Cwiklinski and Donald C. Grant for their technical assistance.

Literature Cited

1. Dent, F. J., Blackburn, W. M., Millett, H. C., *Trans. Inst. Gas Eng.* (1937) **87,** 231.
2. Friedman, S., Kaufman, M. L., Wender, I., *J. Org. Chem.* (1971) **36,** 694.
3. Weller, S., Pelipetz, M. G., *Ind. Eng. Chem.* (1951) **43,** 1243.

4. Dent, F. J., Blackburn, W. H., Millett, H. C., *Trans. Inst. Gas Eng.* (1938) **88**, 150.
5. Walker, P. L., Shelef, M., Anderson, P. A., *Chem. Phys. Carbon* (1968) **4**, 287.
6. Wood, R. E., Hill, G. R., *Amer. Chem. Soc., Div. Fuel Chem., Preprint* **17** (1), Aug. 1972.
7. Feldkirchner, H. L., Johnson, J. L., *Rev. Sci. Instrum.* (1968) **39**, 1227.
8. Blackwood, J. D., McCarthy, D. J., *Aust. J. Chem.* (1966) **19**, 797.
9. Moseley, F., Paterson, D., *J. Inst. Fuel* (1965) **38**, 13.
10. Feldkirchner, H. L., Linden, H. R., *Ind. Eng. Chem.* (1963) **2**, 153.
11. Johnson, J. J., Seminar on Characterization and Characteristics of U.S. Coals for Practical Use, Pennsylvania State University, 1971.
12. Hiteschue, R. W., Friedman, S., Madden, R., *U.S. Bur. Mines, Repts. Invest. No.* **6027** and **6125**, 1962.

RECEIVED May 25, 1973. Work supported by American Gas Association under Contract BR-7S-1.

14

Thermal Hydrogasification of Aromatic Compounds

P. S. VIRK, L. E. CHAMBERS, and H. N. WOEBCKE

Stone & Webster Engineering Corp., P. O. Box 2325, Boston, Mass. 02107

The decomposition of simple aromatic molecules is examined to gain insight into possible reaction pathways involved in coal hydrogasification. Experimental data on benzene and anthracene decomposition kinetics in the literature suggest that the rate-determining step involves destabilization of the aromatic ring. Decomposition rates are substantially independent of hydrogen partial pressures from near zero to 100 atm; over this range the dominant product changes from solid carbon (coke) to methane gas. The associated experimental activation energies are proportional to benzene and anthracene delocalization energies as calculated from Dewar's theory of odd-alternant hydrocarbons. Carbon-forming reactions and the synthesis of aromatic molecules during pyrolysis of paraffinic hydrocarbons are also studied.

In synthesizing low sulfur fuels from coal the Stone & Webster process uses the step-by-step addition of hydrogen to coal under conditions which minimize coke production. The first step involves the conversion of solid coal to a liquid by mild hydrocracking in the presence of a recycle solvent. In the next step these liquids react further with hydrogen under more severe conditions to produce methane, ethane, and aromatics.

To obtain favorable reaction rates the product gas must contain some unreacted hydrogen. The synthesis of ethane makes this possible and at the same time meets the objective of a high volumetric heating value. Aromatic liquids are relatively inexpensive to produce since they contain little more hydrogen than coal. Another advantage is that a Btu of liquid is cheaper to transport by pipeline than a Btu of gas.

The principal reason for gasifying a portion of the liquefied coal is to make substitute pipeline gas which normally is the primary product. Second, as conversion to gas increases, the quality of the residual liquid

Table I. Model Aromatic Molecules

Rings	Name	Structure	Formula	Tb, F
1	Benzene		C_6H_6	176
2	Napthalene		$C_{10}H_8$	424
2	Diphenyl		$C_{12}H_{10}$	491
3	Anthracene		$C_{14}H_{10}$	646
3	Phenanthrene		$C_{14}H_{10}$	643
4	Pyrene		$C_{16}H_{10}$	740
4	Chrysene		$C_{18}H_{12}$	827

improves for use as a fuel. The severe conditions required for hydrogasification reduce the sulfur content of the liquid by-product and improve its transportability and combustion properties. In some special cases it may be ecologically necessary or economically atttractive to convert all of the coal liquid to gas.

The relationships among hydroconversion of the coal liquids, sulfur distribution, and other important fuel properties still need to be evaluated. They are part of the S&W-Gulf development program now under study. This paper summarizes some of the preliminary investigations carried out as a prelude to the development program. Data on reactions by which aromatic molecules are converted to gas are reviewed and correlated; consideration is also given to the formation of aromatic molecules during pyrolysis of hydrocarbons.

Aromatic Decomposition

The main part of this study is limited to decomposition reactions involving small aromatic molecules. The compounds studied are shown in Table I. The largest, chrysene, has a hydrogen content of 5.3 wt % compared with 7.9 wt % for benzene. The detailed chemical pathway(s) for aromatic molecular hydrogenolysis is unknown but it is convenient to consider it as involving three steps: (1) aromatic ring destabilization, (2) breakdown to fragments, and (3) fragment reactions.

Aromatic Ring Destabilization. The above demarcation stems from the well known chemical premise that aromatic compounds owe their unusual stability to a delocalization of pi electrons among the ring molecular framework. For aromatic molecules to react, their delocalization energy must be overcome. Since this energy is large, about 40 kcal/mole, initial destabilization of the aromatic ring is invariably the rate-determining step. This argument, presented in considerable detail by Dewar (1), predicts that the reactivity of all aromatic compounds should be ordered inversely to their delocalization energies (DE); the latter can be computed by Dewar's molecular orbital theory of odd-alternant hydrocarbons.

Table II. Reactivity to Methyl Radical Attack

Compound	Delocalization Energy[a]	Experimental Reaction Rate Relative to Benzene[b]
Benzene	1.155	1
Diphenyl	1.032	5
Naphthalene	0.904	22
Phenanthrene	0.899	27
Chrysene	0.833	58
Pyrene	0.755	125
Anthracene	0.632	820

[a] Nondimensionalized by Hückel factor 2β (see text).
[b] Data from Ref. 2.

Some indication of how theory compares with observation is given in Table II which shows the relative rates at which methyl radicals attack some of the compounds of interest (2). Notice that all rates are ordered inverse to delocalization energies. Values of DE quoted in Table II and hereafter have been nondimensionalized by the Hückel factor 2β; this nondimensional DE is also known as the Dewar number. The pattern of Table II is observed for a variety of other aromatic reactions such as nitration and sulfonation (1). Among the molecules considered in this study (Table I), benzene and anthracene represent the extremes of reactivity.

A second consequence of the rate-determining initial ring destabilization step is that the further course of reaction exerts little influence on the overall rate; therefore a given aromatic molecule should react at a rate essentially independent of the products being formed. This implies, for example, that the rates of benzene decomposition during hydrogenolysis and pyrolysis should be comparable even though the products, methane and coke, respectively, are strikingly different.

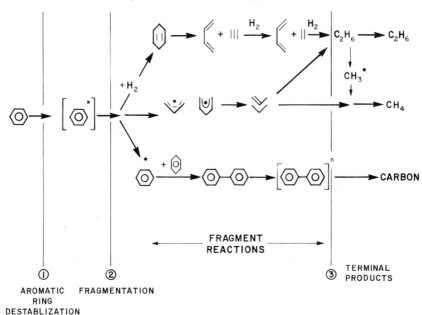

Figure 1. Benzene hydrogenolysis pathways

Breakdown to Fragments. Possible pathways for aromatic decomposition are illustrated in Figure 1. The destabilized aromatic ring is a short-lived species which will either revert to the original stable aromatic ring or break down to various fragments. In the latter event some of the fragments will be nonaromatic and, hence, subject to conventional reaction pathways. For example, the destabilized benzene nucleus may go to cyclohexadiene or it may go to phenyl, pentadienyl, or allyl radicals or to various acetylenes which will further pyrolyze or be hydrogenated. For aromatics with multiple rings like anthracene, the initial breakdown products will likely contain smaller aromatic rings—*e.g.*, benzene—in addition to nonaromatic fragments.

Fragment Reactions. The nonaromatic fragments formed from aromatic ring breakdown can undergo a variety of reactions: (a) molecular reactions such as simple fission (pyrolysis) or hydrogenation–dehydrogenation; (b) concerted electrocyclic reactions, for example, fission and rearrangements; (c) free radical chain reactions such as hydrogenation-dehydrogenation and polymerization.

The complexity of possible fragment pathways can be reduced by certain generalizations.

(1) Molecular fissions have high activation energies about equal to the strength of the bond being broken. As a result, larger hydrocarbons break much faster than the very smallest.

(2) Concerted electrocyclic reactions are faster than molecular reactions which involve separate bond-breaking or -making steps.

(3) Free radical chains, when operative, can be much faster than molecular pathways. At the high temperatures required for hydrogenolysis, free radicals will abound, and it is reasonable to suppose that the hydrogen–olefin–paraffin chain pathways are so fast that equilibrium prevails among these components.

Rate and equilibrium data indicate that the segments of the pathway from benzene fragmentation to ethane formation will be fast relative to benzene destabilization and ethane pyrolysis. Also, whereas the ring destabilization (step 1) is expected to be essentially unaffected by hydrogen, the subsequent product pathways (steps 2 and 3)—whether hydrogenolysis to gas or pyrolysis to coke—should be strongly influenced by hydrogen concentration. Finally, multiple-ring aromatics will break down to both nonaromatic and aromatic fragments; the former will decompose further by the reactions of step 3 while the latter will tend to lose side chains and go to benzene, the stablest aromatic, which will then further react *via* the pathways of Figure 1.

Data Analysis

The hypothesis that aromatic reaction rates are controlled by the ring destabilization step can be tested by comparing the rates of hydrogenolysis and pyrolysis. If true, the rates of decomposition of a given aromatic compound should be identical for either process. Further, reaction rates and their associated activation energies should correlate with the delocalization energy of that compound. Sources of experimental information for the aromatic compounds of interest are listed in Table II along with associated reaction conditions. In each case, the data were processed by the usual methods to yield first-order rate constants (k_1, sec^{-1}) as a function of temperature for the initial decomposition of the aromatic:

$$A \xrightarrow{k_1} \text{products}: \frac{dC_A}{dt} = -k_1 C_A$$

The assumption of first-order kinetics was not generally verifiable from the data and consequently the precision of these inferred rate constants is not especially good but the rates are probably of the right order of magnitude in all cases. Results for benzene and anthracene, theoretically expected to be the extreme cases, are presented in Figure 2.

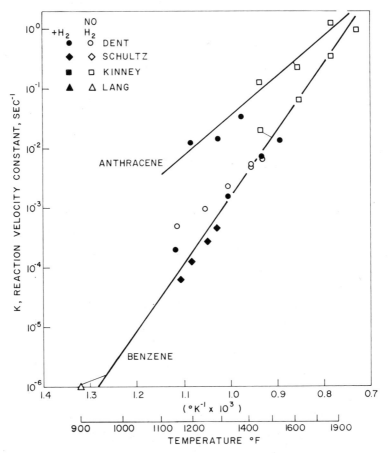

Figure 2. *Rates of decomposition for benzene and anthracene*

Benzene Decomposition Rates. Although the data of separate investigators can each be fitted with straight lines of somewhat different slopes, all the data are adequately described by the single heavy line shown. This indicates that the rates of benzene decomposition during hydrogenolysis and pyrolysis are essentially the same over a wide range of experimental conditions. In particular, the insensitivity to hydrogen pressure, which varies from near zero to 100 atm, is noteworthy. The experimentally observed equality among benzene decomposition rates suggests a common rate-determining step which, in turn, lends support to the thought that aromatic ring destabilization common to both hydrogenolysis and pyrolysis reactions is rate determining.

Benzene Decomposition Products. Further insight into the reaction pathway can be obtained from the reported reaction products. In the presence of substantial hydrogen the lowest temperature data, at 900°F (6), show diphenyl as the sole product whereas the higher temperature data, at 980°–1200°F (4) and 1100°–1500°F (3), indicate mainly methane and some ethane as products. The mole ratio C_2/C_1 tends to unity at benzene conversions below 5% and approaches zero at high benzene conversion. The diphenyl product suggests either a destabilized ring breakdown to a phenyl fragment or a concerted hydrogen elimination from two benzene molecules. It is also interesting because it represents net dehydrogenation of the benzene for purely kinetic reasons even though thermodynamic equilibrium strongly favors gasification.

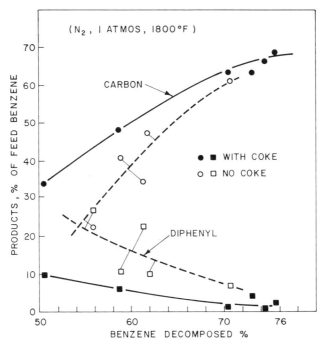

Figure 3. Effect of coke on product distribution for benzene pyrolysis

None of the above authors report coke (carbon) formation nor do they mention any hydrogenated C_6 liquid products. However, hydrogen balances on the data of Schultz and Linden (4) reveal that the empirical formula C_6H_n of the C_6+ components does change from $n = 6$ to $n = 8$

Table III. Summary of Experimental Data for

Compound	Ref.	Type[a]	Temp., °K	Press., atm
Benzene	3	H	873–1123	50
	3	P	873–1073	50
	4	H	800–973	200
	5	P	1073–1373	1
	6	H	758	250
	7	P	1473	1
Diphenyl	5	P	1073–1373	1
	6	H	773	200
Naphthalene	4	H	838–958	200
	5	P	1073–1273	1
Anthracene	3	H	923–1073	50
	5	P	1073–1273	1
Chrysene	5	P	1073–1273	1
	8	H	723	70

[a] H = hydrogenolysis, P = Pyrolysis, B = batch, F = flow.

as benzene conversion proceeds from 0 to 50%, indicating at least some direct hydrogenation of the C_6 ring.

In the absence of much hydrogen (pyrolysis), the gaseous reaction products are principally hydrogen and methane. The H_2/CH_4 mole ratio is variable, about 2–4 in Dent's experiments (3) (1100°–1,450°F, 50 atm N_2) and 8–30 for Kinney and Delbel (5) (1450°–2000°F, 1 atm N_2). Dent (3) also reports small amounts of ethane ($C_1/C_2 = 1$) at 1100°–1300°F while Kinney and Delbel (5) detected traces of acetylene.

Dent (3) does not mention coke or condensed products, but Kinney and Delbel (5) report diphenyl and carbon (coke) as the major products of benzene pyrolysis and show further (Figure 3) that the diphenyl/carbon product ratio decreased in the presence and increased in the absence of coke packing, even though the packing did not appreciably affect the overall benzene decomposition rate. The implications concerning the benzene-to-diphenyl-to-coke pathway are: (1) both ring destabilization and breakdown are probably noncatalytic, homogeneous gas-phase steps and (2) the carbon formation reaction is catalyzed by the product, coke, and probably does not involve further benzene participation. Finally, the very highest temperature data of Kinney and Slysh (7) indicate that the primary benzene decomposition products are hydrogen, acetylene, and diacetylene (solid carbon forms from the latter two); smaller amounts of diphenyl and other C_3 and C_4 acetylenes were also observed. The primary decomposition is interesting because the three main products could stem directly from benzene *via* a concerted pericyclic reaction involving four electron pairs; according to the Woodward-

Aromatic Hydrogenolysis and Pyrolysis

Run Conditions		Reactor	
HC mole fraction	Diluent	Type[a]	Residence, sec
0.10	H_2	F	60
0.10	N_2	F	60
0.15	H_2	B	1000
0.10	N_2	F	2–40
0.35	H_2	B	10^4
0.001–0.01	He	F	0.004–0.112
0.05	N_2	F	2–40
0.05	H_2	B	2×10^4
0.10	H_2	B	1000
0.02	N_2	F	2–40
0.04	H_2	F	60
0.01	N_2	F	1–40
0.005	N_2	F	2–40
0.13	H_2	B	10^5

Hoffman rules (9) such a pathway would be thermally allowed if it involved an antarafacial sigma bond opening.

Anthracene Decomposition. The two sources of anthracene decomposition data are Dent (3) and Kinney and Delbel (5). The coincidence between decomposition rates during pyrolysis and hydrogenolysis of anthracene, like benzene, also supports the notion that ring destabilization is rate determining. Decomposition products from anthracene pyrolysis noted by Kinney and Delbel (5) were mainly carbon, with the carbon formation catalyzed by coke. Product gases were mainly hydrogen and methane, $H_2/CH_4 = 10$, with traces of acetylene. The hydrogenolysis products noted by Dent (3) were mainly methane and ethane and small aromatic rings, benzene, and naphthalene. No carbon formation was reported. Dent (3) reports only the fraction of anthracene converted to gas, but his data suggest that the breakdown of a destabilized anthracene ring in the presence of hydrogen leads to one benzene molecule as a fragment. The gas associated with this initial anthracene breakdown contains methane and ethane in the mole ratios $C_1/C_2 = 3.5$ at 1200°F and 5.3 at 1300°F. This does not yield any clear clues about the nonaromatic fragments except perhaps that a 4-carbon species (which would give $C_1/C_2 = 2$) may be involved. The change in C_1/C_2 ratio with temperature is too large to be explained by simple ethane pyrolysis with methyl radical hydrogenation.

Other Aromatic Molecules. Decomposition rate data for some of the other aromatic molecules of interest calculated at the conditions of Table III are shown in Figure 4. Substantially all of the points lie between the anthracene and benzene limits, and reasonably straight lines can be drawn to represent the variation of decomposition rate constant *vs.* temperature for each of the molecules.

Decomposition products observed were as follows:

Diphenyl. During hydrogenolysis at 930°F and 200 atm H_2, benzene was the sole product. The products of pyrolysis, besides coke, are not clear because the diphenyl results are derived from the benzene pyrolysis data of Kinney and Delbel (5).

Naphthalene. During hydrogenolysis at 1160°F and 200 atm H_2, Schultz and Linden (4) report methane, ethane, and small amounts of propane in the gas with the molal $C_1/C_2 \sim 1$ at low conversions. Benzene

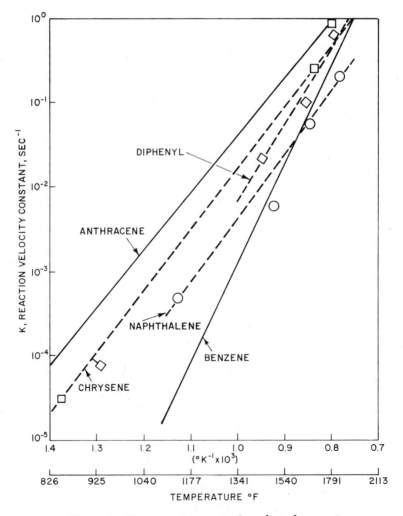

Figure 4. Decomposition rates for selected aromatics

was detected in significant amounts, and traces of toluene and ethylene were also found. Hydrogen balances indicate some direct hydrogenation of the C_{10} ring as well, but no coke formation was reported. During pyrolysis Kinney and Delbel (5) found the gaseous products are mainly hydrogen and methane with traces of acetylene at 1500°–1800°F in N_2 at 1 atm. The principal product was solid carbon, and traces of condensation products like 2-2'-binaphthyl and perilene were also detected. The binaphthyl is analogous to diphenyl and suggests an analogous pathway to coke.

Table IV. Arrhenius Parameters for Aromatic Decomposition Rates

Compound	A, sec^{-1}	E^* kcal/mole	$T_{1/2}(1000°\,K)$,[a] sec
Benzene	4.4×10^8	52.6	499
Diphenyl	1.6×10^7	43.1	118
Naphthalene	4.5×10^5	36.8	171
Chrysene	3.4×10^5	33.5	43
Anthracene	1.8×10^5	30.7	20

[a] $k_1 = A \exp(-E^*/RT)$; $T_{1/2}$ = half-life = $(0.693/k_1)$.

Chrysene. Orlow and Lichatschew (8) found that with 70 atm H_2 the hydrogenolysis reaction products (by weight) were 25% methane, 35% coke, with the remaining 40% containing phenanthrene, naphthalene, benzene, and various hydrides of each. The pyrolysis products were hydrogen and methane with traces of acetylene in the gas and solid carbon.

Correlation

The experimental decomposition rate constant data can be fitted to Arrhenius expressions of the form:

$$k_1 = A \exp(-E^*/RT)$$

for each molecule. Values of these Arrhenius parameters, the pre-exponential factor A and activation energy E^*, are collected in Table IV which also lists for orientation, the corresponding decomposition reaction half-life at 1340°F (1000°K). According to theory the activation energy should be proportional to the delocalization energy—i.e., a plot of E^* (experimental) vs. delocalization energy (calculated) should have all molecules lying on the line between the origin and the benzene coordinates. Figure 5, an arithmetic plot of activation energy vs. delocalization energy, shows a trend in accord with theory. Anthracene and benzene, the two cases with the most data, are in especially good agreement.

Effect of Hydrogen

While it appears that the rate-determining aromatic ring destabilization step is essentially unaffected by hydrogen, the products of decomposition most assuredly are. Increasing hydrogen concentration switches the decomposition pathway from pyrolysis, which leads primarily to solid carbon (coke), to hydrogenolysis, where the product is gas, mainly methane. Understanding how hydrogen concentration controls the crossover between pathways is of interest. However, since the detailed pathway is not explicitly defined, we will focus only on a few aspects expected to be important. Much of the following discussion refers to benzene decomposition because this case has the most data.

Thermodynamic Equilibrium. The equilibrium concentrations of H_2, CH_4, and C_6H_6 are dictated by the following reactions:

	k_p at 1340°F	
	a	b
$C_6H_6 + 9H_2 \rightarrow 6CH_4$	10.6	−4
$C_6H_6 \rightarrow 6C + 3H_2$	16.57	+2
$CH_4 \rightarrow 2H_2 + C$	1.011	1

where a is the exponent to the base 10 and b is the power of the pressure term in atmospheres.

Calculations for this system show that carbon can always form before benzene has reached gasification equilibrium. Further, at atmospheric pressure, carbon formation can occur at very low benzene conversions, unless a very large excess of hydrogen is used. At a fixed hydrogen-to-benzene ratio, increasing the total pressure favors gasification and retards carbon deposition, based on equilibrium considerations.

A study was therefore made of the effect of total pressure, hydrogen-to-benzene ratio in the feed, and benzene decomposition on the gross heating value of the product gas. The study was limited to conditions at which ratios of hydrogen to methane in the product gas would be greater than that required to inhibit the presence of carbon at equilibrium. The results are presented on Figure 6.

At 50% benzene decomposition, the maximum heating value that can be obtained at 1400°F is about 800 Btu/scf while at 1500°F the gross heating value (GHV) would be reduced to about 600 Btu/scf—under conditions where no carbon could exist at equilibrium. The principal curves—*i.e.*, those relating benzene conversion with GHV of product gas—are those for constant pressure and hydrogen-to-benzene ratio in the feed. The H_2/C_6H_6 ratio selected for plotting at a given total pressure was that leading to maximum product GHV for a given benzene conversion.

Carbon Formation

Since coke is the terminal product of the aromatic pyrolysis pathway, it is of interest to explore the formation mechanism. Insight into this process in the range 800° to 1100°C is provided by the benzene pyrolysis data of Kiney and Delbel (5) in a flow reactor. The diphenyl concentration vs. time behavior reported is characteristic of an intermediate in a sequential reaction A → B → C where A (benzene) decreases and C

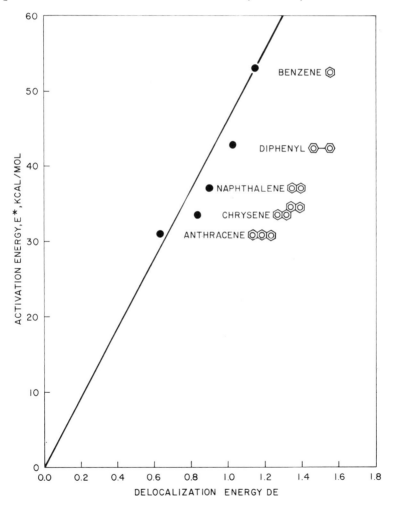

Figure 5. Relation between activation and delocalization energies

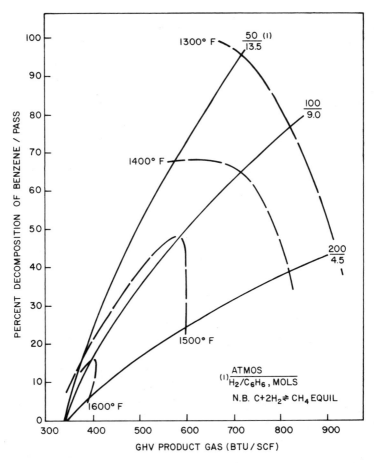

Figure 6. Effect of operating variables on gasification of aromatics

(carbon) increases, both monotonic with time while the intermediate B (diphenyl) increases at small times and decreases at long times. It is also instructive to compare results at the same temperature, 1800°F, with and without coke packing as shown in Figure 3 and discussed earlier. This further suggests that carbon formation proceeds through a sequence of reactions in series

$$\text{C}_6\text{H}_6 \xrightarrow{k_1} \text{C}_6\text{H}_5\text{-C}_6\text{H}_5 \xrightarrow{k_2} \text{carbon} \tag{1}$$

The first reaction is unaffected by coke whereas the second is catalyzed by it. Removal of catalyst would slow down the second reaction, thus increasing the intermediate diphenyl concentration as observed.

In regard to the molecular reactions leading to carbon, the literature contains many references to a benzene-by-benzene addition with hydrogen elimination

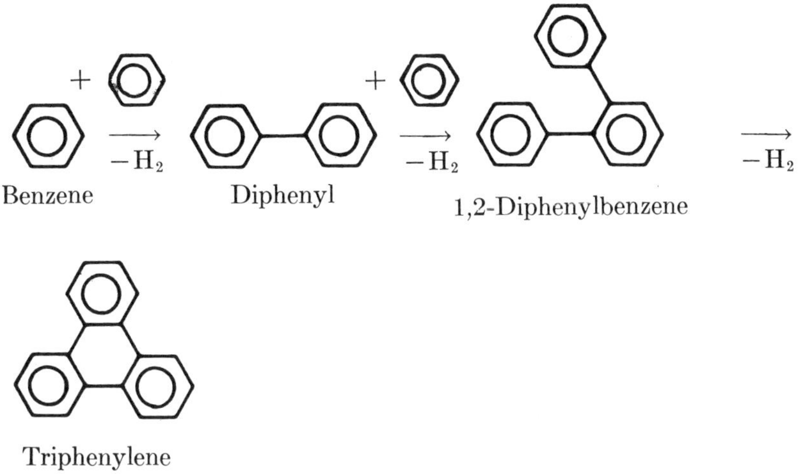

Benzene + Benzene $\xrightarrow{-H_2}$ Diphenyl + Benzene $\xrightarrow{-H_2}$ 1,2-Diphenylbenzene $\xrightarrow{-H_2}$

Triphenylene

(2)

and some of the intermediate products—*e.g.*, diphenylbenzenes and triphenylene, have been detected in the tarry residue resulting from benzene pyrolysis. However if Reaction 2 were the main pathway to carbon, it would essentially involve benzene in every step, so carbon formation should be in very high order in benzene. Catalytic effects enhancing carbon formation should strikingly increase the benzene decomposition rate (and *vice versa*). This is not the case as noted above. Further, reactions with benzene in every step would face the maximum benzene delocalization energy (DE) barrier compared with reactions between more condensed species with less DE than benzene. Thus, although the concentrations of the condensed species would undoubtedly be lower than benzene, the adverse effect of lower concentration on overall reaction rates could easily be offset by the lower activation energies and hence higher rate constants, of the more condensed molecules. A plausible alternative scheme for the main pathway to carbon formation is therefore of the form:

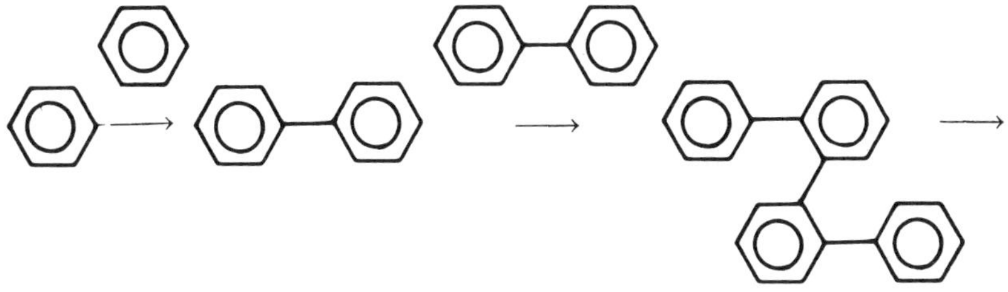

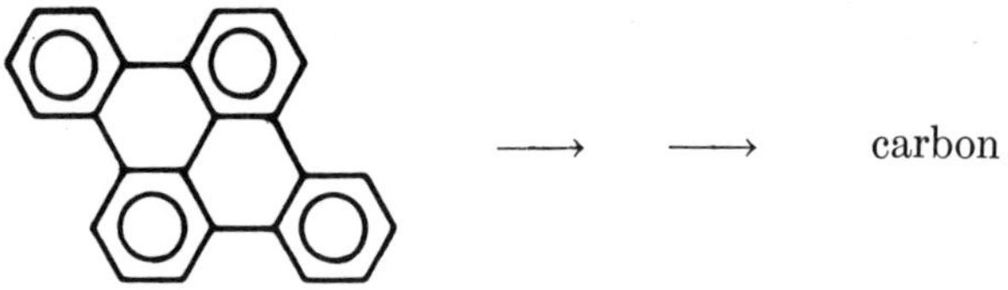

[structure] ⟶ ⟶ carbon   (3)

which involves 1,2,4 ... benzene nuclei rather than the 1,2,3 ... sequence of Reaction 2. According to the above scheme, since the bigger molecules are more reactive, the overall rate should be controlled by the first few steps, namely,

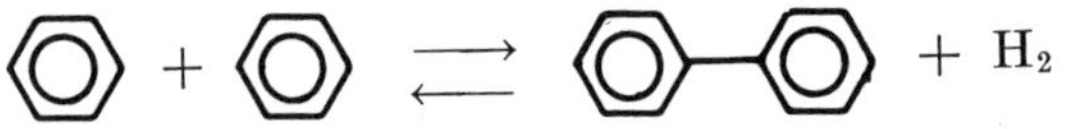

⌬ + ⌬ ⇌ ⌬—⌬ + H_2 (4)

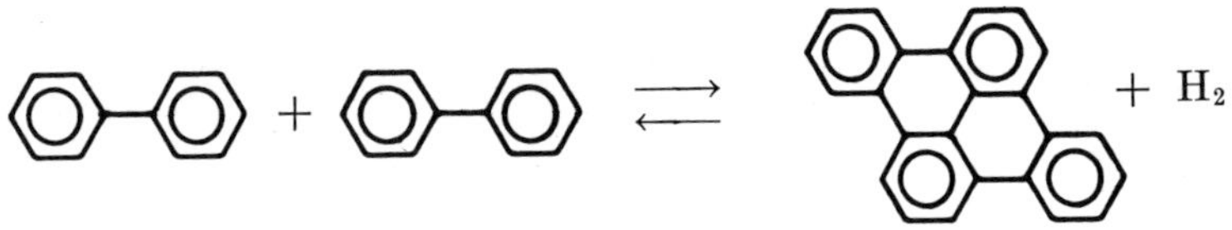

⌬—⌬ + ⌬—⌬ ⇌ [structure] + H_2 (5)

(a and b) [structure] ⇌ [structure] + H_2 ⇌

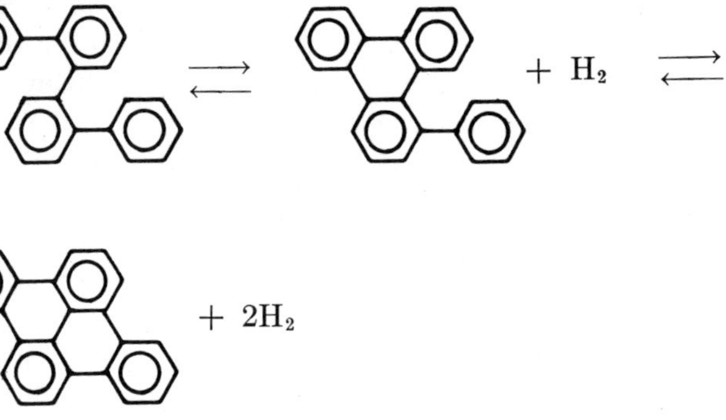

[structure I] + $2H_2$

I

(6)

Reactions 6 (a and b) are intramolecular hydrogen eliminations which one would expect to be fast compared with the bimolecular hydrogen elimination Reactions 4 and 5; therefore the reactants of Reaction 5 can, in effect, be considered to yield the products of Reaction 6b. Now, if we let I approximate carbon, the essential components of the alternative benzene-to-carbon pathway are:

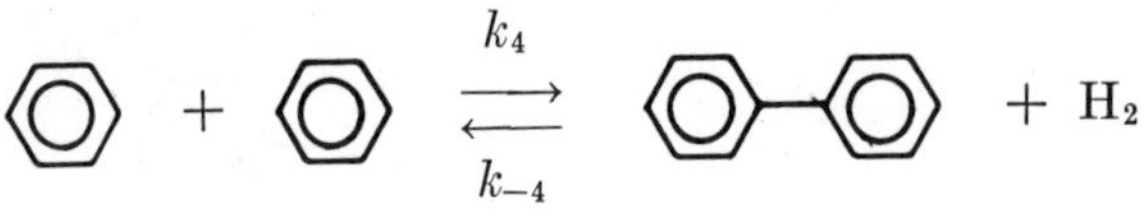

⌬ + ⌬ $\underset{k_{-4}}{\overset{k_4}{\rightleftarrows}}$ ⌬—⌬ + H_2 (4)

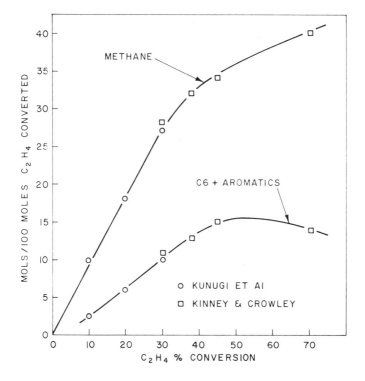

of which Reaction 4 is a homogeneous gas-phase reaction unaffected by coke whereas Reaction 5 can be catalyzed by coke product.

Figure 7. Pyrolysis of ethylene

Data of Kinney and Delbel (5) for benzene pyrolysis to carbon may be modeled by the above scheme of two sequential reactions (4 and 5) simplified such that (a) both reactions are kinetically limited in the forward directions and (b) Reaction 4 is at equilibrium while Reaction 5 is forward-kinetics controlled. Case (b) appears the more plausible for the bulk of the data, but some of the experimental trends at low conversions at the lower temperatures are qualitatively as well predicted by case (a).

Aromatic Synthesis

The foregoing has dealt with aromatic molecule decompositions; a related chemical pathway involves the synthesis of aromatic molecules during pyrolysis and hydrogasification of paraffinic hydrocarbons. The exact mechanism for the thermal synthesis of aromatics from paraffinic or naphthenic molecules is not fully understood although most investigators conclude that it probably involves olefins as an intermediate step. Early investigations showed that aromatic liquids could be produced from all simple olefins and paraffins. Maximum aromatic yields of 5 wt % were obtained from methane by pyrolysis at 1050°C for 10 sec while propane gave a yield of 12% at 850°C. In general olefins were found to give higher yields of aromatic liquids than paraffins. For example, at 10 sec residence time propylene yielded 19% aromatics at 800°C compared with 12% at 850°C already noted for propane.

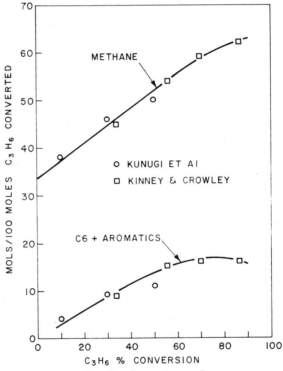

Figure 8. Pyrolysis of propylene

When synthesizing aromatics from paraffins or olefins, molecules containing a higher weight fraction of hydrogen than the feed molecule must also be formed to keep the reaction in hydrogen balance. Figure 7

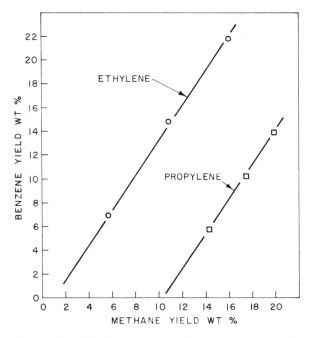

Figure 9. Ethylene and propylene pyrolysis (yield as per cent of feed converted)

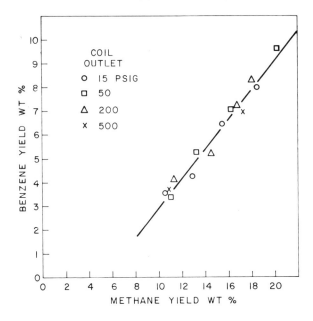

Figure 10. Naphtha pyrolysis

shows data of Kunugi *et al.* (*10*) and Kinney and Crowley (*11*) relating methane and aromatics yield with ethylene conversion. The data of both investigators form a smooth curve, the former limited to low conversions. Figure 8 presents similar information for propylene. For both olefins, aromatics and methane production are linear with conversion up to maximum aromatics production. This is shown more clearly in Figure 9 which is a replot of data from the first two figures.

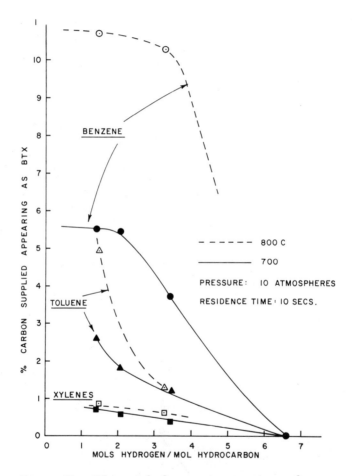

Figure 11. Effect of hydrogen concentration and temperature on aromatics production

Cracking data for naphtha obtained from the Stone & Webster bench-scale pyrolysis unit show this same characteristic relationship (Figure 10). Here, a somewhat unexpected and interesting result is that the relationship between methane and aromatics yield is essentially independent of pressure over a wide range of pressure conditions. This effect of hydrogen on aromatic synthesis can be evaluated from the data of Moignard and Stewart (*12*) (Figure 11) who used an essentially aromatics-free light hydrocarbon feed stock. At 700°C, no aromatics formation was observed when the hydrogen/feed hydrocarbon ratio exceeded 6.6. As the relative amount of hydrogen is reduced, both toluene and benzene syntheses rapidly rise with the benzene tending to an upper limit at about 2 to 3 moles of hydrogen per mole of feed. The same trend is again apparent at 800°C, but here the benzene production is considerably higher, presumably a result of the increased conversion of feed stock. Toluene production decreases rapidly as excess hydrogen increases and is virtually independent of temperature at hydrogen concentrations above 3 moles per mole of feed stock. The decrease in aromatics production is presumably a result of the hydrogenation of the olefinic intermediates, preventing the formation of aromatic molecules.

Conclusion

(1) The rates of decomposition of simple aromatic molecules are essentially independent of hydrogen partial pressure from near zero to about 100 atm. However, increasing hydrogen concentration does change the dominant decomposition product from solid carbon (coke) to methane gas.

(2) The experimentally observed activation energy for the decomposition of an aromatic molecule is linearly related to its delocalization energy as calculated from Dewar's theory.

(3) The formation of carbon from benzene at low temperatures (800°–1100°C) proceeds through diphenyl as an intermediate and probably does not involve any further benzene-by-benzene addition and dehydrogenation.

(4) Equilibrium considerations suggest that the production of a high Btu gas product from benzene without coke formation will require operation at several hundred atmospheres and at relatively low conversions of benzene per pass, using hydrogen partial pressures below those needed for stoichiometric conversion of benzene to methane.

(5) In the absence of hydrogen, aromatics are synthesized during pyrolysis of low molecular weight paraffins and olefins; in both cases olefinic intermediates are probably involved. As the ratio of hydrogen to hydrocarbon is increased, the synthesis of aromatics is inhibited; during decomposition of a light paraffinic naptha at 700°C, no aromatic products were formed when the hydrogen-to-hydrocarbon mole ratio exceeded 6.

Literature Cited

1. Dewar, M. J. S., "The Molecular Orbital Theory of Organic Chemistry," McGraw-Hill, New York, 1969.
2. Levy, M., Swarc, M., *J. Amer. Chem. Soc.* (1955) **77**, 1949.
3. Dent, F. J., *Rept. Joint Res. Com.*, 43rd, British Gas Council, 1939.
4. Schultz, E. B., Linden, M. R., *Ind. Eng. Chem.* (1957) **49**, 2011.
5. Kinney, C. R., Delbel, E., *Ind. Eng. Chem.* (1954) **46**, 548.
6. Lang, K., Hoffman, F., *Brennst.-Chemie* (1929) **10**, 203.
7. Kinney, C. R., Slysh, R. S., *Proc. Carbon Conf.*, 4th, Buffalo, 1957, p. 301 (1960).
8. Orlow, N. A., Lichatschew, N. D., *Chem. Ber.* (1929) **62B**, 719.
9. Woodward, R. B., Hoffman, R., "The Conservation of Orbital Symmetry," Academic, New York, 1971.
10. Kunugi, T., Sakai, T., Soma, K., Sasaki, Y., *Ind. Eng. Chem., Fundam.* (1969) **8**, 374.
11. Kinney, R. E., Crowley, D. J., *Ind. Eng. Chem.* (1954) **46**, 258.
12. Moignard, L. A., Stewart, K. D., *Inst. Gas Eng. Meetg.*, 29th, Nov. 18 and 19, 1958.

RECEIVED May 25, 1973. A portion of this work was done while P. S. Virk was affiliated with the Massachusetts Institute of Technology.

INDEX

INDEX

A

Abstraction reactions	54
Acetylene in various gases, decomposition of	35
Acetylene, isotopic composition of	39
Additives, standard gasification tests for screening	181
Alkali carbonate catalysis of coal–steam gasification	203
Alkali carbonate catalyst, recoverability of	212
Amplitudes, chromatographic deflection	51
Analysis, coal	94
Analysis, Glenrock coal	205
Analyses of coals	115
pretreated coal	181
vitrains, proximate	2
Anthracene decomposition	245
Arc hydrocarbon synthesis	42
process, coal conversion	30
reactor, rotating	33
Area, effect of hot zone	47
Area, hydrocarbon distribution vs. surface	48
Aromatic decomposition	239
rates, Arrhenius parameters for	247
molecular hydrogenolysis	239
molecules, model	238
ring destabilization	239
synthesis	254
Arrhenius parameters for aromatic decomposition rates	247
Arrhenius plot for methane production	65
Atom conditioning time, H-	60
Atomic hydrogen–carbon reaction cell	56
Atomic hydrogen reacting with carbon	54

B

Back-scattering electron micrographs (BSE)	234
Balances, sulfur mass	213
Bed pyrolysis reactor system, fluidized	10
Benzene decomposition products	243
Benzene decomposition rates	242
Benzene hydrogenolysis pathways	240
Bi-Gas process	20
for producing synthesis gas	126
Bituminous coals by rapid heating, devolatilization of	4
Boundary layer, eliminated diffusion	82
Breakdown to fragments	240
Bituminous coal char gasification, kinetics	145

C

Carbon atomic hydrogen reacting with	54
formation	249
gasification rates	185
monoxide and gasification of carbon, increase in production of	189
reaction cell, atomic hydrogen-	56
reaction, steam-	102
target	61
Carbonaceous feed, solid	52
Carbonate catalyst, recoverability of alkali	212
Carbosphere	75
Catalyzed chars, reaction of non-	221
hydrogasification, effect of temperature on non-	225
hydrogasification non-	222
hydrogasification of coal chars	217
Catalysis of coal–steam gasification, alkali carbonate	203
Catalysis of coal–steam gasification, nickel	203
Catalyst(s) multiple	204, 208
Raney nickel	191
recoverability of alkali carbonate	212
stability of sprayed Raney nickel	194
systems, comparison of char-	230
admixed with coal	181
nickel methanation	204
specific rate of gas production using various	185
specific rate gas production with various	182
Catalytic methanation	207
Cathode (FCC), fluid convection	43
Cell, atomic hydrogen–carbon reaction	56
Cenosphere	75
Chains, free radical	241

261

Chemistry, gasification	128
Char(s)	
catalyzed hydrogasification of	127
catalyst systems, comparison of	230
combustion	72
composition of air-pretreated hvab Pittsburgh No. 8 coal	149
gasification, kinetics of bituminous coal	145
microscopy of	232
reaction of non-catalyzed	221
Chromatographic deflection amplitudes	51
Coal	
analyses of pretreated	181
analysis	94
char(s)	
composition of air-pretreated hvab Pittsburgh No. 8	149
gasification, kinetics of bituminous	145
catalyzed hydrogasification of	217
conversion arc process	30
conversion process, NRRI	207
devolatilization	92
gasification	
catalysis at elevated pressure	179
entrained	126
scheme, schematic of	220
by rapid heating, devolatilization of	1
by rapid heating, devolatilization of subbituminous	5
Glenrock analysis	205
hydrogasification of raw	110
plasma pyrolysis of	29
pyrolysis, mathematical model of	16
pyrolysis model, criteria for a	14
ratio, oxygen	107
reactor outlet temperature vs. oxygen fed/lb	95
steam gasification, alkali carbonate catalysis of	203
steam gasification, nickel catalysis of	203
Cold finger, liquid helium	55
Comparison of char-catalyst systems	230
Combustion	
equivalence ratio	104
flame, pyrolysis	75
heterogeneous	72
Composition(s)	
of air-pretreated hvab Pittsburgh No. 8 coal char	149
hot zone surface	48
and production rates, pyrolysis gas	14
gas stream	22
Concentration, effect of hydrogen	100
Conditioning time, H-atom	60
Conditions on overall plant efficiency, effect of gas purification	25
Convection cathode (FCC), fluid	43
Conversion arc process, coal	30
Conversion process, NRRI coal	207
Criteria for a coal pyrolysis model	14

D

Decomposition	
of acetylene in various gases	35
anthracene	245
aromatic	239
methane	131
in the PEDU, methane	132
products, benzene	243
rates, Arrhenius parameters for aromatic	247
rates, benzene	242
Deflection amplitudes, chromatographic	51
Delocalization energies (DE)	239
Destabilization, aromatic ring	239
Devolatilization	150
of bituminous coals by rapid heating	4
coal	92
of coal by rapid heating	1
of subbituminous coal by rapid heating	5
Dilute-phase (FCP) reactor, free-fall	109
Dilute-phase reactor	108
Diffusion boundary layer, eliminated	82
Diffusion effects, pore	83
Distribution vs. surface area, hydrocarbon	48

E

Electron micrographs (BSE), back scattering	234
Eliminated diffusion boundary layer	82
Energies (DE), delocalization	239
Entrained coal gasification	126
Entrained two-stage gasifier	127
Equipment development unit (PEDU), process and	128
Equivalence ratio, combustion	104
Ethane as a function of temperature, production of	64

F

Feed rate variables	94
Feed, solid carbonaceous	52
Finger, liquid helium cold	55
Flame, pyrolysis-combustion	75
Flow rates, material	22
Fluid bed gasification	160
Fluidized-bed pyrolysis reactor system	10
Fluid convection cathode (FCC)	43
Flyash particulates, solid	9
Fragments, breakdown to	240
Free-fall, dilute-phase (FDP) reactor	109

INDEX

Free radical chains 241
Free radical recombination
   processes 54

G

Gas
   Bi-Gas process for producing
      synthesis 126
   compositions and production
      rates, pyrolysis 14
   and oil, synthetic 72
   pipeline-quality 116
   production
      influence of K_2CO_3 on methane and total 209
      of high Btu................ 115
      with various catalysts, specific rate 182, 185
   purification conditions on overall plant efficiency, effect of .. 25
   shift reaction, water 116
   (SNG), synthetic natural 207
   stream compositions 22
   turbine inlet temperature on overall plant efficiency, effect of 25
   with a solid surface, reaction of 80
Gases, temperature profiles of 6
Gasification
   alkali carbonate catalysis of coal steam 203
   of carbon, increase in production of carbon monoxide and .. 189
   chemistry 128
   correlations of low-rate 158
   entrained coal 126
   fluid-bed 160
   in hydrogen, effect of temperature on low-rate 160
   in hydrogen, rate constants for low-rate 168
   kinetics of bituminous coal char 145
   low-rate 150
   in steam-hydrogen mixtures, effect of temperature on low-rate 161
   in steam, effect of temperature on low-rate 159
   nickel catalysis of coal–steam .. 203
   physics 137
   rates, carbon 185
   rates for catalyst systems 209
   schematic of coal 220
   steam–oxygen 108
   tests for screening additives, standard 181
Gasifier, entrained two-stage 127
Gasifier system, Synthane 198
Glenrock coal analysis 205

H

Heating, devolatilization of
   bituminous coals by rapid 4
   coal by rapid 1
   subbituminous coal by rapid .. 5

Helium cold finger, liquid 55
Heterogeneous combustion 72
High Btu gas, production of 115
High pressure thermobalance 146
Higher volatile matter loss 76
Hot zone area, effect of 47
Hydrane process 108
Hydrocarbon distribution vs.
   surface area 48
Hydrocarbon synthesis, arc 42
Hydrogasification, effect of temperature on non-catalyzed .. 225
Hydrogasification
   of coal chars, catalyzed 217
   non-catalyzed 222
   of raw coal 110
   pressurized 108
   thermal 237
Hydrogen
   abstraction reactions 54
   concentration, effect of 100
   effect of temperature on low-rate gasification in 160
   increase in the production of methane and 188
   mixtures, effect of temperature on low rate gasification in steam 161
   rate constants for low-rate gasification in 168
   reacting with carbon, atomic .. 54
   yield, increased 199
Hydrogenolysis, aromatic molecular 239
Hydrogenolysis pathways, benzene 240

I

Inlet temperature on overall plant efficiency, effect of gas turbine 25
Interior thermocouple temperature 122
Isotopic composition of acetylene 39

K

K_2CO_3 on methane and total gas production, influence of 209
Kimber 75
Kinetic models 148
Kinetics of bituminous coal char gasification 145

L

Liquid helium cold finger 55
Loss, higher volatile matter 76
Low-rate gasification 150
   correlations for 158
   in hydrogen, rare constants for 168
   in hydrogen, effect of temperature on 160
   in steam, effect of temperature 159
   in steam–hydrogen mixtures, effect of temperature on ... 161

M

Mass balances, sulfur 213
Mass transfer 80

Material flow rates	22
Matter loss, higher volatile	76
Mathematical model of coal pyrolysis	16
Methanation catalysts, nickel	204
Methanation, catalytic	207
Methane	
decomposition	131
in the PEDU	132
and hydrogen, increase in the production of	188
formation, rapid rate	150, 157
formation stage, correlations for rapid-rate	152
as a function of temperature, production of	64
production, Arrhenius plot for	65
production vs. residence time, optimum temperature for	135
and total gas production, influence of K_2CO_3 on	209
yield, residence time and temperature effect on	134
yield, increased	200
Micrographs (BSE), back-scattering electron	234
Microscopy of chars	232
Model aromatic molecules	238
Model, criteria for a coal pyrolysis	14
Models, kinetic	148
Molecular hydrogenolysis	239
Molecules, model aromatic	238
Multiple catalyst	204, 208

N

Natural gas (SNG), synthetic	207
Nickel	
catalyst	
of coal–steam gasification	203
Raney	191
stability of sprayed Raney	194
methanation catalysts	204
NRRI coal conversion process	207

O

Oil, synthetic gas and	72
Outlet temperature vs. oxygen fed/lb coal, reactor	95
Oxygen	
coal ratio	107
fed/lb coal, reactor outlet temperature vs.	95
gasification, steam–	108

P

Particle temperature	86
Particulates, solid flyash	9
Pathways, benzene hydrogenolysis	240
PEDU, methane decomposition in the	132
Phase reactor, dilute–	108
Phenomena, surface-controlled	67
Physics, gasification	137
Pipeline-quality gas	116

Plant efficiency, effect of gas purification conditions on overall	25
turbine inlet temperature on overall	25
Plasma pyrolysis of coal	29
Pore diffusion effects	83
Pressure and gas composition on low-rate gasification in steam–hydrogen mixtures, effect of	162
Pressurized hydrogasification	108
Pretreated coal, analyses of	181
Process and equipment development unit (PEDU)	128
Process(es)	
coal conversion arc	30
Hydrane	108
NRRI coal conversion	207
for producing synthesis gas, Bi-Gas	126
free radical recombination	54
Synthane	1
Production rates, pyrolysis gas compositions and	14
Production vs. residence time, optimum temperature for methane	135
Products, benzene decomposition	243
Profiles of gases, temperature	6
Propane as a function of temperature, production of	65
Properties of solids, thermophysical	2
Pulverized coal	72
Purification conditions on overall plant efficiency, effect of gas	25
Pyrolysis	72
combustion flame	75
gas compositions and production rates	14
increased volatiles from rapid	4
mathematical model of coal	16
model, criteria for a coal	14
of coal, plasma	29
reactor system, fluidized-bed	10

Q

Q factor	75

R

Radical chains, free	241
Raney nickel catalyst	191
stability of sprayed	194
Rapid	
heating, devolatilization of bituminous coals by	4
coal by	1
subbituminous coal by	5
Rapid-rate methane formation	150, 157
correlations for	152
Rapid pyrolysis, increased volatiles from	4
Rate(s)	
Arrhenius parameters for aromatic decomposition	247
benzene decomposition	242
carbon gasification	185

INDEX 265

Rate(s) *(continued)*
constants for low rate gasification
in hydrogen 168
effect of temperature and steam 191
gas production with various
catalysts, specific 182
of gas production using various
catalysts, specific 185
material flow 22
for non- single-, and multiple-
catalyst systems 209
pyrolysis gas compositions and
production 14
variables, feed 94
Ratio, combustion equivalence ... 104
Ratio, oxygen/coal 107
Raw coal, hydrogasification of ... 110
Reaction(s)
cell, atomic hydrogen-carbon .. 56
fragment 240
of a gas with solid surface 80
hydrogen-abstraction 54
of non-catalyzed chars 221
products, production of volatile 3
shift 104
steam-carbon 102
water-gas shift 116
Reactor
dilute-phase 108
free-fall, dilute-phase (FDP) .. 109
outlet temperatures vs. oxygen
fed/lb coal 95
rotating arc 33
size, effect of varying 99
system, fluidized-bed pyrolysis .. 10
thermobalance 147
wall temperature 122
Recombination processes, free
radical- 54
Recoverability of alkali carbonate
catalyst 212
Residence time 99
optimum temperature for methane production vs. 135
and temperature effect on
methane yield 134
Ring destabilization, aromatic ... 239
Rotating arc reactor 33

S

Scattering electron micrographs
(BSE), back 234
Schematic of coal gasification ... 220
Screening additives, standard gasifi-
cation tests for 181
Shift reaction 104
water-gas 116
Sintering, excessive 201
Solid
carbonaceous feed 52
flyash particulates 9
surface, reaction of a gas with a 80
Solids, thermophysical properties of 2

Specific rate gas production with
various catalysts 182, 185
Stage gasifier, entrained two- ... 127
Standard gasification tests for
screening additives 181
Steam
–carbon reaction 102
effect of temperature on low-rate
gasification in 159
gasification, alkali carbonate
catalysis of coal- 203
gasification, nickel catalysis of
coal- 203
hydrogen mixtures, effect of temperature on low-rate gasification in 161
oxygen gasification 108
rate, effect of temperature and .. 191
Stream compositions, gas 22
Subbituminous coal by rapid heating, devolatilization of 5
Sulfur mass balances 213
Surface
arc, hydrocarbon distribution vs. 48
composition, hot zone 48
controlled phenomena 67
Synthane gasifier system 198
Synthane process 1
Synthesis
arc hydrocarbon 42
aromatic 254
gas, Bi-Gas process for producing 126
Synthetic gas and oil 72
Synthetic natural gas (SNG) ... 207

T

Target, carbon 61
Temperature
effect on methane yield, residence time and 134
interior thermocouple 122
on low-rate gasification in hydrogen, effect of 160
on low gasification in steam,
effect of 159
on low-rate gasification in steam-
hydrogen mixtures, effect of 161
for methane production vs. residence time, optimum 135
on non-catalyzed hydrogasification, effect of 225
on overall plant efficiency, effect
of gas turbine inlet 25
particle 86
production of
ethane as a function of 64
methane as a function of 64
propane as a function of 65
profiles of gases 6
and steam rate, effect of 191
Time
H-atom conditioning 60
optimum temperature for methane production vs. residence 135

Time *(continued)*
   residence 99
   and temperature effect on methane yield, residence 134
Tests for screening additives, standard gasification 181
Thermal hydrogasification 237
Thermobalance, high pressure 146
Thermobalance reactor 147
Thermocouple temperature, interior 122
Thermodynamic equilibrium 248
Thermophysical properties of solids 2
Turbine inlet temperature on overall plant efficiency, effect of gas 25

V

Variables, feed rate 94
Vitrains, proximate analyses of ... 2

Volatile matter loss, higher 76
Volatile reaction products, production of 3
Volatiles from rapid pyrolysis, increased 4

W

Wall temperature, reactor 122
Water-gas shift reaction 116

X

X-ray micrographs 234

Z

Zone area, effect of hot 47
Zone surface composition, hot 48

The text of this book is set in 10 point Caledonia with two points of leading. The chapter numerals are set in 30 point Garamond; the chapter titles are set in 18 point Garamond Bold.

The book is printed offset on Danforth 550 Machine Blue White text, 50-pound. The cover is Joanna Book Binding blue linen.

*Jacket design by Norman Favin.
Editing and production by Spencer Lockson.*

The book was composed by the Mills-Frizell-Evans Co., Baltimore, Md., printed and bound by The Maple Press Co., York, Pa.